Environmentalism and Politics

Volume 3

Fax
Messages from a near future

Full list of titles in the set
ENVIRONMENTALISM AND POLITICS

Fax
Messages from a near future

Jorge Wilheim

publishing for a sustainable future

London • New York

First published in 1994

This edition first published in 2009 by Earthscan

Copyright © 1996 by Jorge Wilheim

ISBN 978-1-84971-004-6 (hbk Volume 3)
ISBN 978-0-415-84763-6 (pbk Volume 3)
ISBN 978-1-84971-001-5 (Environmentalism and Politics set)
ISBN 978-1-84407-930-8 (Earthscan Library Collection)

For a full list of publications please contact:

Earthscan
2 Park Square, Milton Park, Abingdon, Oxon OX14 4RN
Simultaneously published in the USA and Canada by Earthscan
711 Third Avenue, New York, NY 10017
Earthscan is an imprint of the Taylor & Francis Group, an informa business

First issued in paperback 2013

Earthscan publishes in association with the International Institute for
Environment and Development

A catalogue record for this book is available from the British Library

Library of Congress Cataloging-in-Publication Data has been applied for

Publisher's note
The publisher has made every effort to ensure the quality of this reprint, but
points out that some imperfections in the original copies may be apparent.

At Earthscan we strive to minimize our environmental impacts and carbon
footprint through reducing waste, recycling and offsetting our CO_2
emissions, including those created through publication of this book.

Dear Professor J.

I am a cultural historian, and I am preparing my masterplus
on the process of technological acculturation during the
thirty years of the Transitional Trauma...

In the mid-1990s, Professor J. receives a series of faxes from a
researcher known only as Titus, who, through a displacement in the
space–time continuum, is writing from the year 2025.

During the course of their remarkable corresondence, Titus
tells J. of the technological advances which will shape urban
development in the future, and of how contemporary societies and
institutions will be transformed following a phase of violent unrest in
the late 20th century. J. learns of the 'big jam' – a pile-up of 14,500
cars which bought about the demise of personal transport (and later
became the site of a popular theme park); the dramatic fall of the giant
Space Tower (a particularly stupid 'intelligent building'); Viavita, a
biological sewage treatment; and the basic ingredients of a good 21st
century pizza.

FAX

Messages from a

Near Future

FAX

Messages from a

Near Future

Jorge Wilheim

Translated from the Portuguese by
Kate Clayton

EARTHSCAN

Earthscan Publications Ltd, London

First published in Portuguese in 1994 by Editora Paz E Terra SA, Brazil
Translated from the Portuguese by Kate Clayton

First published in English in 1996 by Earthscan Publications Limited, London

A catalogue record for this book is available from the British Library

ISBN 1 85383 376 2

Page design by S&W Design
Cover design by Dan Mercer, Power FX

For a catalogue, or more information, contact
Earthscan Publications 120 Pentonville Road London N1 9JN
Tel: (44) 171 278 0433
Fax: (44) 171 278 1142
Email: earthsales@earthscan.co.uk

Earthscan Publications Limited is an editorially independent subsidiary of Kogan Page
Limited and publishes in association with the International Institute for Environment
and Development and WWF-UK

Foreword

With more than half of the world's rapidly growing population expected to live in cities by the beginning of the new millennium, the demand on urban design is daunting. But the potential is equally great. The changes in building design, transportation, sanitation, communications in every aspect of urban life will revolutionise the city.

Jorge Wilheim introduces us to the changes that society must implement, and – as the man in charge of the UN conference on the future of the city, Habitat II – he is uniquely well-placed to know what is needed. But he also has the wisdom and imagination to understand that making cities humanly sustainable is not only a matter of policy, but also of individual commitments. What matters, and the point of striving for sustainability, is that we as individuals have so much to gain from it.

This short book is a revelation and an inspiration. Using the simple but brilliant device of an exchange of faxes from across time, Jorge reveals to us what the future may hold, and reduces to the inevitable the critical decisions which society should be taking today to protect it. And he does so with charm, brio and a very engaging sense of fun.

Sir Richard Rogers

Foreword

Is it really impossible to foresee the near future, in this transitional period of history we are living through? After all, it is being formed now, just in front of our eyes ... Being part of this process, too near to it, we have difficulties in focusing; we might be looking, but we do so without seeing.

Imagination and intuition might help. We can surf on this ocean of uncertainty and feel the breeze of times to come. Relax. Let me tell you a story...

During the last Saturday of Carnival I had an extraordinary experience: I received and replied to a series of faxes whose content was nothing less than staggering. Bewildered but enriched by what I had learned, uncertain how to convey what had happened to me, I discussed a number of the topics with friends before taking the decision to hand over the following narrative to a publisher.

The physicist José Zats taught me something about the relativity of time, the economist Fernão Bracher discussed certain aspects of banking with me, the socio-economist Ignacy Sachs clarified a number of matters and I debated various hypotheses with the sociologist Manuel Castells. I should like to thank them here for the time they spent discussing the doubts aroused by the content of these faxes – and I hereby absolve them of any and all responsibility for what follows! I went through all the fax messages once more when preparing this English translation, and some lines were added or corrected.

I am extremely grateful to Vera Severo for all her valuable input and for her painstaking critical analysis of a narrative almost beyond belief.

I thank Kate Clayton for her enthusiasm in translating and the editors, Jo O'Driscoll and Caroline Richmond, for their careful searching (and finding) of inconsistencies to be explained

Finally, I also thank Joanna Wilheim for her constant encouragement and for reading so constructively the various versions which it proved necessary to set down.

9:00

Carnival. Saturday morning. The ruby flowers of the bougainvillaea on the low wall of the terrace outside my office were trembling in the morning breeze as I reorganised the piles of paper on my desk which the cleaners – with disciplined obstinacy – disorganised on a daily basis. I had remained in São Paulo in splendid isolation, in order to finish some work.

It was at this moment that the fax machine broke the silence, with a shrill but discreet tinkling to be heard coming from behind my armchair. My reaction, as always, was one of enjoyable anticipation. It was rather like the pleasure I feel arriving home if I catch sight from a distance of the brightly-coloured tip of an air-mail envelope from abroad, or the stamp-covered bulk of the latest issue of a review to which I subscribe.

This enjoyable anticipation of news, of surprises, is the exact opposite of the painful suspense I feel when I'm listening to messages recorded on the answering machine – strange voices firing rapid communiqués for which I'm never ready, with even the most familiar voices distorted (out of sheer perversity) just to worry me.

After the fax gives its incoming message tone there's usually a pause while you wait – generally in some frustration – as the message retreats and peeks out and leaves you there holding your hand out. A pause of mounting suspense, until at last you hear that productive purring from the innards of the machine as the mechanism finally begins to run, announcing that the unknown is about to materialise.

1

And that morning brought neither the frustration of being cut off nor that of receiving one of those mail-shots which brazenly take advantage of a communications channel left open in the generous expectation of receiving news. A white tip emerged, like a tongue glimpsed between its owner's lips, and rapidly turned into a sheet of paper printed with text pouring out upside-down. Curious, I immediately looked for some identification, a heading ... nothing. I had to wait for the end of the message – a single sheet – to cut the page with the impatient and clumsy movement that usually means I end up tearing the corner.

And I read:

> Dear Professor J.:
>
> Although we do not know one another, I am taking the initiative of asking for your collaboration. I obtained your name and fax from the "Anthropos XX" database, together with the thematic reference index which indicates that you have written papers and even books in your own language on the period I am researching.
>
> I am a cultural historian and I am preparing my masterplus on the processes of technological acculturation during the thirty years of the Transitional Trauma. The titles of your papers suggest that you have studied similar subjects in your own country, so I hope, sir, that you will excuse my troubling you with a few enquiries. If you do not have time to answer them, perhaps you could suggest another scholar or some backup source on this period. I have had considerable difficulty retrieving it, and constant use of electronic translation leads to the loss of nuances which could shed a great deal of light on the subject, think you not?
>
> Scanning my outline I would say that the first questions are simple: do you happen to know in which year the

privatisation of the telegraph and post office was completed in your country? Do telephone apparatuses still exist, and, if so, what is the proportion between these antique communicators and the combination type (phone/faxvideo)? You may find it surprising that I do not have this information about your country, but although our institutions are generally well-informed there is a shortage of information about Brazil in the period I am researching.

Thank you very much in advance for any information you may send me.

In greatest solidarity,

your colleague Titus (0423, 95-EU 07/240)

9:15

Who the hell was this Titus? Some practical joker? His name looked a bit odd to me ... and his Portuguese was so pedantic, correct enough but full of strange neologisms. The text read like a translation, and had several odd mannerisms I couldn't put my finger on. I glanced at the automatic record at the top of the fax in order at least to identify where it had come from: 427GMW411222. Too many digits! I searched in vain through the telephone directory but I couldn't identify the country. Maybe interference had caused some of the digits to be repeated? Or maybe there was some kind of fault in his machine. I would ask Titus himself where he was writing from. I set aside the problem of identification for the moment and turned my attention to the content of his enquiries.

Well of course our government has never even considered including the National Post among the state companies for privatisation. And anyway, who would be prepared to buy it with so many incumbent state employees? Especially now, just when it looks likely that all written communications are going to shift to electronic media or fax and be sent by satellite? A letter addressed to someone in your own city takes a few seconds by fax instead of 48 hours in the post ... from São Paulo to London, twenty seconds instead of the usual ten days! There's no way the post can compete in efficiency. Either the National Post changes its branch of activity or it goes bankrupt. It won't be long before stamps are sold only in antique shops – collectors take note – and at extremely high prices, since they'll be that rare

5

when the postal service has disappeared completely!

But where on earth could a historian so entirely ignorant of our daily lives hail from? This had to be some young man in love with his own powers of invention, rewriting history after his own fashion in order to make it fit in with some pre-invented theoretical blueprint. It wouldn't be the first time ...

And what could this "Transitional Trauma" be? Perhaps he was referring ironically, or with typical Nordic humour, to the turn of the millennium? If so, the thirty years he's researching would be the period from 1970 to 2000: we haven't even got there yet, and already here's a historian wanting to study how Brazil copes with the great changes these years are bringing!

Still, it's true to say that the period he's chosen is one rich in significance. It was preceded by an important decade: the brilliant 60s, with the Cold War, man reaching the moon, the hippie movement and the student unrest of '68. That ushered in the 70s, with the Vietnam War, the military dictatorships in Latin America and economic growth fuelled by petrodollars: the decade of OPEC and the search for alternative energy sources, as well as the decade of terrorism and the increasingly consolidated strength of the transnational companies. A historian would find the 80s unusually interesting too, because of Gorbachev's perestroika, the fall of the Berlin Wall, the weakening of the nation states, AIDs, the success of the environmental movements, the rise in cocaine trafficking and consumption, the trend towards Neo-Liberalism: a list of topics that would interest any scholar. Or perhaps Titus wanted to study the origins underlying what came next, the break-up of the Soviet Union and the end of ideological bipolarity? The last decade of the millennium, still a mystery – but certainly likely to be traumatic. Perhaps even, as some people fear, the last decade before Nostradamus' terrible prophecy is fulfilled ...

Undeniably, these are thirty years worth studying. But I had never heard of anyone referring to them as a "Transitional Trauma". Could I have understood correctly? On the other hand I thought Titus said something about acculturation ... I read the fax again: technological acculturation. Was he interested in local adaptations of

6

universal technologies, or in the adaptation of particular customs to new technology? I replied with a long and detailed fax, describing the government's privatisation policy and elaborating on the dim prospects facing our National Post. At one point I speculated:

> Surely the distribution structure that this institution has perfected over the decades in most countries, covering the whole of each national territory, could be put to good use in some way or another. But what could it distribute? Perhaps your typical postal service could become an efficient bulk distributor within cities, with its thousands of delivery staff working out of sorting centres. So messages would get distributed by electronic means while bulky items were distributed by delivery staff. Or maybe the postal service companies could turn themselves into networks of local communications centres, selling communications services and marketing small communications-related products.

Although I answered Titus' enquiry quickly, I was actually more interested in finding out who this was on the other end of the line than in discussing what was on the agenda. I added a string of impatient questions and settled down to wait rather anxiously for another fax to bring me more news.

Barely twenty minutes later I was rewarded with a long fax pouring onto the ground, like ectoplasm streaming out of someone's nose. Titus thanked me for my prompt reply but showed a certain puzzlement:

> You describe as a hypothesis something that has existed for more than twenty years! Although I do not know your country, I do nevertheless have sufficient information to

7

know that, after the long gap which separated our countries and particularly after the stagnation during which we did not receive sufficient explanations on what was happening there, the technological differences of thirty years ago no longer exist. To judge by your answer I would have to believe that Brazil is most backward! Yet I know that it is not; there must have been a widening of regional differences, and perhaps, sir, you are referring to the poor region of your continent? I remain perplexed ... Such a pity that I cannot read and write your language: perhaps the translator in my equipment leads you to misunderstand my enquiries. A shame, think you not? ... Of course all messages are transmitted by electronic means; how could it be otherwise? Although there has been a gradual change in the supports used, since television signals are for the most part transmitted by optical fibres, while telephony increasingly uses satellite transmission. Do you recall the year in which every citizen in the so-called "modern" world received his code, thus becoming part of the global information and communications technology network? What chaos! And how people complained when, after making frivolous juvenile calls, they found the cost duly debited to their accounts. Did you get used to carrying your MC with you (do you wear it on your belt, on your wrist, as a necklace or in your pocket)? I personally detest carrying the microcontactor, because it gives me the feeling of being vulnerable to importunate calls from anywhere in the universe! Oh God, such pretentiousness, you will say ... But after the dramatic vicissitudes we have gone through, after the extraordinary increase in the density of our cities, I have ended up – like so many other citizens – valuing privacy in the extreme. I leave my MC at home. The fax, however, does not form part of my dislikes. It would be impossible to exchange messages with efficiency without it. Even the email does not work for me as a mail-box – I forget to respond to its signal of "incoming

8

messages"! Especially now that we have finally overcome some inconveniences such as the lack of secrecy, the interference of unwanted advertising messages, the occurrence of transmission faults, the absence of colour, the imperfection of photographs. Nowadays the Korean, Chinese and Singaporean models are extremely satisfactory, as well as cheap. What model do you use?

As you probably know, people use faxes a great deal in public cars here in the Community these days. Although faxes have been a common optional extra in other cars for years, they were not normally installed in public cars for fear of vandalism or theft. Now they have codes inserted in them which autodestruct any equipment removed from the car. I have already seen this autodestruction occur: it is impressive! But of course we are no longer living in the period of the wars of intolerance, and I expect that vandalism, like all destruction, is a thing of the past in Brazil too. At least for some time ...

But allow me to return to the period of my research and proceed with the enquiries I prepared in my outline. Could you tell me exactly when Brazil introduced the system of public cars in its cities? Do they still use biomass energy (cane alcohol, I think) or were they adapted to use hydrogen from the start? I imagine that the solar battery so well developed here in the Community has come to be a success in Brazil, with all the sun that shines there ... It is incredible that it should have taken almost 170 years to invent and develop a viable battery, small and light, superior to the famous battery of Planchon, is it not? ... Perhaps at last we will use electric cars instead of the hydrogen ones: sun instead of water!

The contents of this fax left me puzzled, and so did the perfectly natural air with which Titus spoke about phenomena which frankly seemed more like science fiction. On the other hand, he hadn't answered my enquiries about the country he was living in. True, he did mention a "Community". But which one? Could it be the British Commonwealth, meaning that he lived somewhere in the former empire? Or was this some community or other within a particular society? Or perhaps he was referring to the European Union? In any case I was beginning to feel a bit sceptical, even though I was curious. It could turn out to be a joke on the part of some friend or other ...

Anyhow, eager to help with his research, I immediately answered the questions about fuel and rather pedantically went into detail about the history of the use of alcohol as a fuel from its beginnings in 1975, when the technology was perfected by Brazilian aeronautics engineers; I explained how they simply gave away this technology to the big manufacturers without even charging the royalties that would have given them the opportunity to go on with their research. I described how difficult it had been to establish a programme to combat the air pollution caused by vehicle emissions, and outlined the relative success achieved in replacing diesel fuel with natural gas in buses and trucks. And I also mentioned the fact that this experiment was ignored by the developed countries for decades so as not to disturb the universe of the oil lobbies ...

In my reply I also enquired about the use of hydrogen as a fuel, as all I knew was that Mazda and Saab were researching its use and that they were promising to have an experimental prototype on the road by the end of the century. I had no idea that there were hydrogen-powered vehicles in use somewhere already. But what really intrigued me was this "public car". It must be some kind of taxi – or was it just a mis-translation? But this time the answer took longer than the twenty minutes I had waited for Titus' first reply.

11.00

I received another message at last. Titus apologised for not having been able to answer right away as he had wanted to, but he had to attend a "prearranged international conference" in the next room. Once more he regretted having to resort to electronic translation, which was certainly imperfect. He noted that although our exchange was of the greatest interest it was incomprehensible in places, leaving him in doubt as to my meaning.

In answer to my enquiry about public cars, he told me:

> I do not understand your question. Our public cars are the same as yours: a small car, a two-seater but with enough space for luggage, packages or bags of shopping. They are quite basic, but have the minimum requirements for urban transport and communications: phone, fax and electronic vehicle location sensing and route planning. We usually pick one up at any service station, where there are always some around (if not, we wait for someone to return one); the car is activated by inserting a credit card, which can be withdrawn from the appropriate slot only when the vehicle is returned to some station after use. Our bank account is then debited, but it is always better to examine the bank duplicate because bank errors are frequent and not likely to be in our favour. Is the Brazilian system the same, or have you adopted linkage

11

to the underground car-park system as in the United States? I can never forget how this all began: the change of habits caused by the Big Jam! Have you already visited the Park? My children visited it during the last school holidays. In fact, as far as the use of the car is concerned, everything changed at that memorable Christmas in 2004 after the great earthquakes ...

I leapt out of my seat! Numb and sweating, I went on reading where I stood:

Do you remember? Until then, from what my parents have told me, people were so much in love with their own car that they took it with them adoringly wherever they went; if they possibly could, they would have kept their car in the drawing room of their house!

The day before my children left on their trip, I had to tell them in detail how several earth tremors had put some of the motorways out of use in the area around Los Angeles that Christmas; how the resulting traffic jam was so huge, with so many accidents, such fighting and unrest, that the pile of vehicles was simply never unscrambled because it proved to be cheaper to claim the insurance and abandon the car! They could not believe it! But just imagine, I told them, there were 14,500 cars and 837 trucks involved in this traffic jam! There were people who stayed in their car for four days, fed by scouts mobilised especially for the occasion. Of course they called in the National Guard to maintain order. But what good did that do? It was total hysteria, because nobody dared to abandon their beloved car! Fourteen women went into labour and 320 senior citizens had to be hospitalised after being picked up by helicopter. As a matter

12

of fact, there were helicopters flying constantly over the whole area — about 4000 hectares, wasn't it? I saw a documentary on the Big Jam recently, commemorating the twentieth anniversary. I was an adolescent then, and I remember the television news perfectly. It was a sight out of Dante, but in the end it was tragicomic — everybody shouting, rushing around, fighting over a crumpled wing when really there was nothing that could be done about it any more: there had been over 300 collisions, making it impossible to remove the vehicles and restore the flow of traffic!

My children were a great success when they recounted all these details to their classmates when they set out together to visit the Big Jam Ethnological Park, as so many schools do — they go there when they have run out of visits to the various Disneylands scattered across the five continents. By the way, have you been to the Disney-under-the-waves off Havana? But I digress ... anyway, my children were thrilled with the trip. And apart from the fact that of course it is amusing to find mountains of cars from a bygone age, with brightly coloured wings and gleaming bonnets sticking up unexpectedly out of the ground among hills and meadows, I think the Big Jam Park is significant for the alteration it brought to our habits and to urban transport, think you not?

The incident had huge repercussions all over the world, and everybody inevitably took part in the debate it provoked, which revealed what it should have been easy to understand decades ago. A moving car occupies 30m², and if we multiply this value by the desired number of cars (one for each adult) in our metropolises the resulting area would not be compatible with that of the roads actually available for traffic. It was logically impossible to realise Henry Ford's so-called American dream, and the closer we got to it the more the traffic ground to a halt and people's nerves could not take it. I had to explain to my children how we arrived at the current strategy not long ago — and not without first having to overcome the resistance of manufacturers, unions and a good many users: as far as the individual car is concerned, we must optimise use, not maximise quantity.

Oh God! Sometimes it is so difficult to explain certain things to adolescents: they could hardly believe that the use of the automobile had changed just because of the Big Jam. In fact, it is hard for them to accept that the course of history is so often made by twists of fate, changed by accident or for reasons of passion. Sir, do your children also look always for rationality, for cause and effect? At times, in my scepticism, I feel so very old ...

Anyway, the event of the Big Jam had such resounding repercussions that it forced people to think, accelerated the development of collective transport systems with financial support from the insurance companies and made us revise our habits of using the individual car. Cars started being designed and produced in a diversified form suited to these new habits: the family car for travelling, and in the city the taxi and the public car which you so wondered at. The taxi, obviously, is far and away the most comfortable manner of circulating in the metropolis — and yet driving a public car alone preserves that old pleasure and feeling of freedom, of independence, which was always associated with the automobile from its earliest beginnings. Think you not?

I sat down again, dumbfounded. Titus was writing to me from another time: he spoke of the year 2004 as the past! Could he be aware that we were out of phase like this? I interrupted my reading and flung myself at my Toshiba to send him the following note of explanation scrawled with a shaking hand:

Titus: something very strange is happening to us. To be perfectly clear, I have to tell you that I'm living in 1994! And you – can you be living in 2024?

The answer came back immediately:

Extraordinary, my dear J., extraordinary! This explains some of the ambiguities of our exchange. Our dialogue is diachronous, how beautiful! ... I do not want to waste much time investigating the diachronism because we must make the most of these circumstances and intensify our exchange: we have so much to learn, as long as this situation lasts. Oh God how beautiful! But I cannot help speculating a little: this phenomenon of ours must prove Hawking's old theories, when he describes the nature of the segment in the spherical geometry of time – even though he himself doubted his own theory at one point. During my research into the past, I must somehow have displaced myself towards a temporal pole. As I continued forwards, when I passed that point I actually started to go backwards "on the other side" until I met up with the temporal parallel in the past in which you are living!

We must make the most of this wonderful living example of the relativity of time and answer all that we can. I have no end of questions to ask! I can just imagine how eager you must be. May I drop the formalities and just call you J.? Imagine my bewilderment – and how lucky I am, a historian conversing with the past! Oh how beautiful! Think you not?

In greatest solidarity,

your friend Titus

11:50

My mind was racing so much that I had to step outside and walk round the block in an attempt to order my thoughts. Nothing doing, though! There simply was no way to order the phenomenon I was experiencing. I tried to speculate for myself: in addition to Hawking's theory, the relativity of time itself suggested a possible explanation for this meeting across time planes. If I remember rightly, time can be considered as a mass, and the "present moment" is no more than a slice cut through this mass as if it were a jelly; the plane of the present moment is continuously displaced along an axis perpendicular to it – the time axis. Somehow or other, I thought to myself, when Titus sent his message via the communications satellite the waves encountered some "concavity" in this mass and the resulting refraction caused them to leave the plane on which he lives and impinge on another, earlier plane – the plane of my present moment. Beats me how, though!

All I could do right now was stay in contact. I was more interested in making the most of this strange experience than in finding a rational explanation for it – a conversation with the future! I began to hurry; what was I doing out there in the street when I should have been hurling myself at the Toshiba to shower Titus with questions?!

I didn't want to share the secret with anybody just yet. Besides, if I did I was only going to look ridiculous unless I could overcome their disbelief ... and in any case, all my friends were spending

Carnival at the beach or in the country. I slowed down in doubt as I approached the office; what if Titus' faxes were really a huge practical joke organised by some friend to pull my leg? But who could the joker be? No name suggested itself. And anyway, if the whole thing was a leg-pull I would take my hat off to whoever had the imagination to come up with it!

I went back into the office and resumed my "diachronous exchange" with my new friend. Impatiently I started writing by hand, asking him about his education, his family, the office he worked in. I wanted to form a precise image in my mind of the man I was talking to, and his surroundings. For my part, I described my family and my office. I didn't wait long for an answer:

> The two rooms I occupy at the institute are fairly typical of these buildings of up to 15 floors constructed in the last few decades round the outskirts of the former villages here in the Community; this Germanic region in particular has turned into a pleasant urbanised region with urban nuclei separated by forests. The rooms of my office, as I was saying, are fairly typical: two modules of 20m² with infrastructure points, giving ample scope for the individual arrangement of doors, windows, terraces, aquaria, plants and energy and communications supply points. In my own case I ended up creating a rough reproduction of a childhood memory: my grandfather's study in his country house. I have several Persian carpets on a wooden floor, wood-panelled walls with an ancient cuckoo-clock – a local craft – and a series of prints of seventeenth-century Italian landscapes, heavy floral print curtains, a big desk with a great many drawers, a comfortable cushioned armchair and the communications modules with their displays. The second room is intended for meetings and group work, and has a huge circular table with leather chairs; all the necessary screens for teleconferences are incorporated into the surrounding walls, on which we

18

can also draw, create graphics or put up notices. The computer in this room is of greater capacity than the one in the communications module in the first, allowing faster access and extensive filing; the multimedia system common to all computers is displayed on the wall screens.

From the outside, there is nothing about the institute building to suggest that its internal spaces have all the individuality demanded by today's market. This building has little in common with the autocontrolled buildings of the beginning of the century; its style attempts to recapture the serenity and symmetry of the Renaissance palaces even though everything is prefabricated down to the last detail. Have I given you some idea?

As for my personal life, I must admit I was taken aback: your curiosity is well meant and certainly appropriate in the strange circumstances in which we find ourselves, but nowadays we are not in the habit of asking questions or answering hastily about personal matters. Privacy is such a precious possession, think you not? However, as I have said, the present situation fully justifies our mutual curiosity.

I was born in Bonn, took my degree in anthropology in Berlin and went back for retraining in Princeton and Beijing. The masterplus I am currently preparing will be debated in Jerusalem because the dates are convenient, but in any case it is registered with all the universities affiliated to Anthropos XXI. I am married to Fatma, a doctor whose family is of Palestinian cultural origin, and we have a couple of children, Gure and Mara.

Titus went on in this fax to take the opportunity to ask about the Brazilian buildings of my time, even though this was not part of the list of questions he had prepared before. He asked me why the Brazilian architects of the 1960s stopped using brise-soleil, which had

19

been so successful in the past (suggested by Le Corbusier almost a hundred years earlier), and replaced them with the glass facade. Specifically, he asked me whether the glass used was of some special kind which blocked or filtered the sun's rays, making it more suitable than the simple brise-soleil ...

What could I say? Of course the brise-soleil avoids excessive heat and light by stopping the sun's rays from reaching the glass. Without this architectural device, the only thing you can do in a tropical country is resort to expensive electronic shading of the glass. Or air-conditioning, with all the resulting expenditure of energy – a fad among the clientele, before becoming a fad of the architects themselves. And in Brazil's case there is a certain degree of cultural colonisation to take into account ... Talk about stupid! I preferred not to answer ...

Agreeing that our best bet was to keep asking questions, I asked Titus how things stood with the so-called intelligent buildings in his country. His answer came speeding back:

My dear and distant J.,

Here in the Community people used to talk about autocontrolled buildings; would that be the same thing? I think the adjective "intelligent" should be applied to the project or the architect or the client – why attribute it to a building? Though I really should not be surprised; only yesterday I read an article broaching the subject in a respectable American journal. Now I come to think about it, it might be of interest to you since you cannot have read it. It starts with a banner headline: "Twenty years ago – collapse of the world's biggest intelligent building". Then the subheading: "250 dead and almost 1000 wounded in the greatest tragedy of the century". Do not be impressed by the obvious dramatic exaggeration of the phrase. Unfortunately other, greater tragedies did occur. But the headline rightly reflects the huge impact caused by this disaster, which happened, if I am not mistaken, in Hong Kong.

It was a long time before people knew why the building had collapsed, from an engineering point of view, because the usual corporate spirit ensured that responsibility was shunted for ages from one to another of those who had taken part in its construction. But everything suggests – and I am no engineer – that it was because they placed too much confidence in the automatic monitoring and correction system for movable load distribution. Something failed to register, a beam broke, a slab at the top of the building fell, and the whole building collapsed like a pack of cards.

The long-term result of this disaster, according to the author of this article, was self-criticism in architectural and engineering circles; as well as cautiously amending the technical standards, they reviewed the whole concept of the "autocontrolled building" (which apparently in Brazil you used to call "intelligent"). But you are not alone in your concern about this matter ... The century still began, according to this author, in a climate of irritating arrogance: no professional who wanted to progress in his career dared deny that the buildings he designed were "intelligent". To judge by the real estate advertisements which appeared at the time, some of which I have kept to illustrate the text of my masterplus, these buildings promised to "think" of everything: they controlled climatic conditions, the landscape visible from the windows, water and energy consumption; they provided constant information about what was going on within their structures, gave daily statistics about the number of passengers who had used the lifts (leading, as I discovered, to daily bets in the London pubs which duly became an integral part of British gambling systems) and mapped out the routes taken by each visitor to the building. It even seems that, under the pretext of

keeping tabs on everything, some of these buildings had
really voyeuristic monitoring systems with cameras invading
the privacy of their occupants! A filthy trick, think you not?

Ah ... I don't know what the architecture of the twenty-first century
will be like, but I can certainly see how plausible Titus' description is
today. In a lot of buildings, all this monitoring paraphernalia simply
produces huge piles of reports that get filed away by a bored building
manager, who doesn't pay them much attention and is just annoyed
with having to pay high monthly charges to the company selling the
monitoring services imposed by a paranoid desire for "security", for
having everything under control. The monitoring systems that
pretentiously call themselves the "brain of the building" will end up
filling the same role as the tyrannical Parisian concièrges of the
middle of the century: nothing but snoops and police snouts.
     This fax closed the subject as follows:

It is not fashionable to talk about "autocontrolled buildings"
– or "intelligent" as you call them – any more these days;
the craze is over. After ecology was all the rage at the turn of
the century, which led to buildings being sold more on the
strength of the landscape (natural or artificial) to be seen
from their windows, artificial gardens and aquaria in every
corner of their niggardly little flats, the current trend – to
judge by the daily flood of advertisements in the media – is
for flexible, uncluttered spaces, diversity, multiple planes
and, generally speaking, a critique of uniformity.

Of course I was starting to relish the whole text in quite another way
now. Even talking about a subject as banal as building construction,
the fact that this information was coming to me from "another time"

22

gave it unusual interest and importance. Besides, what I could learn about the future of architecture and engineering brought up the wider subject of the relationship between technology and common sense.

What was going to happen over "my" next thirty years (I was already adopting the period covered by Titus' masterplus as the time-frame for my conjectures) to the six areas of technology that some authors now consider to be at the cutting edge: biotechnology, information and communications technology, microelectronics, space sciences, nanotechnology and robotics? I sent a fax asking Titus how much common sense had been displayed in the development of these sectors.

15:15

And back came my answer at once:

> You are worrying about non-substantial issues; the main
> importance of technological innovations does not lie in
> whether or not they lack common sense but in their capacity
> for producing infinitely more than was asked of them! But I
> will confine myself to answering your enquiry.
>
> In much the same way as in the field of engineering and
> architecture, an analogous process occurred in the six
> sectors you have picked out; a measure of self-criticism
> gradually emerged, strongly motivated by the consumer. In
> the case of biotechnology, much ingenuity was exercised at
> first on futile (though extremely profitable) applications
> such as the production of incredible and fantastical aquaria.
> Later, however, this field generated not only innumerable
> products to be found on every housewife's shopping-list but
> one in particular which has lasted; it has had successive
> improvements under various different names, and I will
> describe its qualities in greater detail by way of example.
> This is the well-known cleaning product VIVAVITA, advertised
> on the world's TV screens in such charming fashion by – I
> think – a Brazilian presenter. I expect that this toilet

cleanser is marketed under the same name in your country, is it not?

The use of this product is really part of the introduction of new technology; the process involved, which is so simple, totally eliminates any need for those vast sewers that had to be slowly and laboriously tunnelled throughout the city. The process (and the product) have become so much part of our everyday lives that cultural historians wonder why there was so much resistance when biological technology was adopted in urban sewerage in the last century.

The relationship between technology and common sense can perhaps be better analysed in the second of the sectors at the cutting edge which you mention: ICT. The combination of information technology and telecommunications, which became increasingly characteristic of the last three decades of the last century, gave rise to a whole series of objects of doubtful usefulness but which were seized upon voraciously by a credulous consumer market eager for novelties. Oh God, I remember, for example, the case of the video-phone: it enjoyed initial success and sold well, but users gradually came to realise how inconvenient it could be having an unexpected voyeur in their home or on their desk; people often forgot to turn them off, which led to all kinds of embarrassing situations, with moments of intimacy or collusion caught on-screen. There was that famous time when a scantily clad president's wife carelessly answered the trilling of a videophone in someone else's home ... In the end, once the novelty had worn off, videophones were used only by the lovesick and impatient (and the lovesick are always impatient, is it not so?) and grandparents who quite rightly wanted to gaze upon the faces of their babbling grandchildren. The market remained limited, even though lovesickness sets in during early adolescence and the life-expectancy of the elderly has increased ...

26

At the opposite extreme – that of sense and usefulness – we witnessed the enormous expansion of the electronic information networks, particularly the so-called AINs (Added Information Networks) which add data and corrections themselves, constantly updating the information they store. To mention only the oldest of these ICT networks, by the end of the century the French Minitel already provided interactive links to over five million homes and businesses, and today it offers its users the most practical of services, such as travel reservations, ticket and hotel bookings, and even supplies the latest figures for economic indicators and complex banking operations all over the world. As companies became increasingly transnational in nature, spreading their activities across the globe, they also created their own networks to improve efficiency and competitiveness, which both depend on the appropriate use of information. Of course, while ICT became a mundane everyday requirement, it also made all of us dependent on keeping these networks working properly – and on their (preferably) ethical use. In this respect who can forget the polemic aroused by two incidents at the beginning of the century: a programme was dreamed up on the interlinked air travel reservations network which biased it in favour of a particular airline, and there was an epidemic of electronic viruses created by political terrorists which almost wiped out the world meteorological information network, greatly hindering the preventative measures so important in agricultural production.

Much as the cutting-edge technology of biotechnology and ICT was eventually channelled towards common-sense uses, so too those working in microelectronics had to choose between the fascination of futile novelties on the one hand – which nevertheless sold well simply because they were novelties – and essential everyday uses on the other;

27

common sense, in other words. Nowadays, with CD-MM (originally called CD-ROM, I believe) everywhere, nobody makes those heavy video-cassette machines called VCRs any more. They may have had 148 functions, but they were only ever actually used to project fuzzy images on tiny screens. Of course nobody really used most of those functions! And yet, even so, the mere fact of advertising such a wide range of possibilities constituted a marketing advantage, think you not? The computers my children use to do their homework are limited to modules for the essential functions only, and are extremely simple to use. Nowadays people buy toddlers' and children's computers suitable for each age-range, like clothes.

Now I come to think of it, though, I suppose I am being over-optimistic: although the little VCDs (Video Compact Discs) we have these days are perfectly adequate for us to watch shows with satisfactory image-quality on the large screens in our rooms, it is difficult to justify the recent launch of games in poor taste such as the electronic "Wipe Out your Neighbours", probably a hangover from the unpleasant period of apartheid at the end of the century. So we should not boast too much about how we show such common sense these days in our use of technology ...

Ach! I must of course mention the era of the now forbidden virtual reality games that alienated so many young people twenty years ago. They preferred the refuge of sensuous daydreaming or violent sadism to facing the unpleasant and difficult reality of the Transitional Trauma. Of course up to the time when we controlled AIDS it was better to let youngsters masturbate with their bulging eyes fixed on their headset screens. But the industry produced increasingly pornographic and sadistic fantasies, and these games turned out to be really dangerous.

The example of the old VCRs having unnecessary functions reminds me of the case of the home computer. These days, after going all through the toddlers' and children's series I mentioned earlier, when we reach adolescence we get a basic computation unit from our parents. Starting with this, we buy or rent only the modules we need for any particular task – available from any neighbourhood electronics supermarket – which considerably brings down the price and reduces both the complexity of our computer and its maintenance requirements.

Forgive me the superficiality of the above information, but I am only an anthropologist, and it is Saturday, after all; if we have the time I will look for more detailed information in the next few days. In any case I must repeat my warning: the rapid advance of technology has of course enabled us to perfect certain products, but this is not where its greatest importance lies, nor its greatest impact on the unfolding of events. Forgive my frankness, but in your own case you are so fascinated with new products that you completely overlook the essentials in this matter.

I have to go back into the next room to see to the Istanbul group, who are currently in conference. But do not break off the flow of communications; I shall expect to hear from you and will be back in a moment.

In solidarity,

Titus

16:00

I spent some time looking up information and composing a long fax in which I tried to answer what Titus had said. Videotext is the nearest thing we have right now to a computerised information service in the home such as the French Minitel network, but it is a very long way indeed from offering the range of uses described by my correspondent.

I agreed with him in that I've hardly ever met anyone (of my own generation, at least) who has actually read the manual – or, rather, the collection of manuals – that comes with a computer; at the very most they read the summary in the User's Guide or look at the Quick Reference section when they're stuck (which is often enough).

I couldn't remember much about the current IT network situation, and I didn't have anyone to hand who could sketch it out for me or tell me how many users the networks cover. Most networks are more and more likely to be private ones, created by economic sectors, by corporations or by organisations which don't explain clearly how they work and take damn good care to maintain secrecy about their data and their users.

I appended a hand-written note to my fax, explaining my apprehensions: the black box of IT is getting steadily bigger and blacker. Paradoxically, the more computerised our society has become – or rather, the more it has invented simplified technology in order to become computerised – the greater the power some people have to manipulate others. In a world where large organisations (both public and private) are getting more dynamic and more active by the day, there is less and less room for the individual to act – just when computerisation theoretically offers greater and greater scope for individual action.

Where is this trend taking us? Towards an Orwellian society, or towards a dangerous flash-point of revolution against control, to be followed by lawlessness and destruction? To what extent could the Internet, as an open and decentralised network, be an answer to these threats?

My fax to Titus ended as follows:

> But all this rapid technological development at the end of the century – when you come right down to it, where does its real impact lie? Isn't it in the fact that some countries end up so far behind others, as far as the modernity of their means of production are concerned?

I was surprised to get an answer right away, in the form of a short fax:

> I glanced in on my way to the toilet to see if there was any message from you. I read your fax quickly, and I am just taking a moment to send you a short, partial answer. The technological innovations of the end of the century increased productivity well above the real demand for products. It took us some time to level out this situation, because the imbalance could only really be overcome when most of the enormous marginalised population of each country eventually gained access to the market. There were very high levels of structural unemployment during the transition period; as early as the last decade of the twentieth century there were already 70 million legal or clandestine

immigrants and 20 million unemployed here in the Community (over 10 per cent of the workforce). This was a very serious factor in fuelling every kind of protectionism, prejudice and hatred of our neighbours, as well as in "justifying" the violence used in defence of privilege – which increased enormously, as you may well imagine.

This was one consequence of increased productivity and the expanding gap between economies, but not the only one; in addition to this, countries actually flaunted their very production system itself – assembling elements manufactured in different countries abroad – and showed off the speed with which technology was developing and changing in order to underline the difference between one country and another: between the fast-moving central economies on the one hand and the slow, marginal ones on the other. Scientists and technologists should not have all the blame for this laid at their door; the real importance of the technological innovations made towards the end of the century lay not in the mere use of ingenious new products but rather in this gap between economies.

I will write about the other matters later, especially with regard to the Cosmonet (which was formerly known as the Internet). I have to go back to the teleconference; the co-ordinator, in Istanbul, has already remarked on my impatience and the fact that I have been leaving the room so often! And the Mexican delegate, who is a most crude person, laughed and asked if I was hiding a woman behind that door!

17:00

Titus really must have been under pressure, what with our exchange of information and his teleconference. What could the subject and significance of his meeting be? There were too many questions to ask ... I was afraid I would get caught up in secondary enquiries and find myself taken by surprise when we eventually lost contact. But in circumstances like these, how can you tell what's essential and what's of secondary importance? I decided to clear up one point so as to confirm or correct some of my hypotheses about certain geopolitical prospects:

> Ever since we made contact you keep mentioning Istanbul. I'm intrigued. By coincidence, there was a debate a few days ago about the destiny of Brazil in relation to the future of those countries we used to call the "Third World" which made particular reference to Turkey: some of those present argued that Turkey is going to play a significant part in political developments over the next few decades. They based their argument on its geopolitical position between the growing expanse of the Islamic nations and the European Union (if the EU ever really succeeds in becoming a community; the only thing to be seen so far has been waves of furious nationalism and fratricidal wars).
>
> The range of the Islamic nations will certainly be swelled by the addition of the breakaway southern republics from the

former Soviet Union. And everything suggests that the more radical aspects of Islamic religious fundamentalism will be adopted by a number of African countries (in addition to North Africa) because – among other reasons – an autocratic leadership could make use of such authoritarian fundamentalism, which subsumes the individual, in order to overcome the ancient and insoluble tribal and ethnic rivalries already existing within those countries.

Furthermore, we should bear in mind that some of the Islamic countries probably possess nuclear arms and that everything points to the existence of a flourishing clandestine arms trade. I ask myself (and I admit I'm afraid to ask you) if the world will see a terrible figure arise: the "blue-turbaned son of Mahomet" of whom Nostradamus speaks in his prophecies about the destruction of the "ancient cities".

These events are already going on behind the scenes, but these days I'm aware of political stirrings in Turkey which may either bring about a civil war due to their inflexibilty and pride, or, on the contrary, result in a period of economic progress. There are two million Turkish emigrants who will be returning from Germany and other European countries, taking back with them four things acquired abroad which will play an important part in new developments there: money, knowledge of other languages, personal links with the economic agents and organisations of the EU and, lastly, technological know-how, whether specific or general – in other words, the knowledge needed to modernise the organisation of production and render it competitive. Plus Turkey has a history of conquest and a proven capacity for assimilating other cultures – as well as an army that's always jumpy ...

Do my speculations have anything to do with the Istanbul group meeting being held in the next room?

18:30

Titus began his answer by explaining that he was taking advantage of a break for individual discussions in the ongoing teleconference to re-read the fax I had sent him and quickly scribble down some further comments:

Dear J.,

Let me start by clarifying a little the question of the documentation I am using for my masterplus. You wonder at the absence of certain data on Brazil which you consider common knowledge. But just think: I am so far away (in time) and much of what you consider common in your everyday life may have been considered trivial during the following decades so that the necessary care was not taken to record and preserve it. And I must admit that up to the end of the last century we in the developed countries were vastly ignorant about Latin America. Think you not?

Nevertheless, I admit that for the period I am researching, particularly the end of the last century and the beginning of this one, there was not much coverage of Brazilian events in the world press, except for the usual sensationalised reporting of such tragic events as rapes, massacres, fires, floods etc. But in this field, unfortunately, our own local news was taking up more and more space ... Strangely, there is much more information on Brazil for the period before 1990 and after 2010. Have you any explanation for this phenomenon?

I answered almost angrily, in just a few lines:

We were lost, Titus, incapable of considered thought or of imagining a possible, desirable future scenario to strive towards and build! We were so worried about the evils of inflation, of corruption and political bickering that we couldn't see the wood for the trees!

Our ruling elite were blind and too comfortably settled into their lordly way of life: when you come down to it, the astronomical rate of inflation (over 1 per cent a day) came in very handy to make the yawning gap between rich and poor seem to vanish by the sleight-of-hand of indexing everything; until we began to control inflation during the mid-1990s, the tensions generated were more or less contained by means of individual stop-gap measures taken by a government eternally in debt and which compensated by printing more and more paper money. Just imagine: the difference between the richest 20 per cent and the poorest 20 per cent of the population was of the order of 26 times! In Sweden the ratio is 5 times. It reminds me of something Bacon said: "Wealth is like manure; it has to be well spread out not to smell foul"!

These short messages forced me to think again about the reasons for Brazil and the rest of Latin America being out of the news for two decades. It cannot be just a matter of language in this increasingly anglophone world. Will it really take us so long to get our stagnant economy moving again? Or is there going to be an economic decline so severe that it puts the continent in the irrelevant position of a Fourth World country? Are we running the risk of slipping "from mere dependency into irrelevance", as Castells put it? Will we mark time, without innovation of any kind, in a state of paralysing dependence on the technological innovations of the First World? Will we stop being competitive? Will the First World stop importing our natural resources and products? Or will the ever-widening gap between rich and poor provoke a real crisis – that is, a crisis of revolutionary change?

19:15

A few minutes later I was surprised to receive yet another message:

Dear J.,

Allow me to turn to another matter. Because my research is oriented more towards anthropological, cultural concerns, the documentation I use is not based on the great changes — the ones in the history books or the electronic documentary archives. These would explain things in terms of broad outlines, giving a referential framework. But my research is on everyday life within the context of these changes.

This is why I sought initially to make a careful reading of Brazilian newspapers, particularly that daily which resulted from the merger of a humorous paper with the most serious newspaper published in São Paulo. I forget the names for the moment and I have not time to consult my databases; I cannot retrieve them with my MC because, as I told you, I left it at home!

As a matter of fact this merger was extremely representative of the new way of interpreting daily events, think you not?

41

The serious newspaper contained such a quantity of facts that although one supposedly read it as a daily it actually took much more than a day to read. Inevitably people found this frustrating. And it seems that its political forecasts reached by logical analysis rarely hit the mark. The humorous paper, on the other hand, was completely intuitive but got quite a few of its forecasts right. In my country, of course, our intellectual tradition leads us to mistrust intuition which eludes rationality and does not confine itself to the facts. But striving towards an understanding of phenomena by means of an intuitive approach is a scientific method which enjoys considerable success in the institute in which I work. The paradigm of this method can be expressed as follows: information plus intuition equals comprehension.

I turned this over in my mind: was the *Estadão* – something like the *Washington Post* – going to merge with the satirical *Pasquim*, or was it going to be our *Financial Times*-type paper, the *Gazeta Mercantil*, that merged with the *Private Eye*-style *Casseta Popular*? And I made a note to ask Titus what institute this was. He went on:

You asked me what VIVAVITA is (or will be). This is a green-coloured liquid which is connected to the flushing mechanism of every toilet, even in water-recycling and water-saving systems – the "grey water systems" where the water used to flush the toilet comes from the showers and washbasins.

The product (it gets given different names on other continents to make it look as if there is competition) is nothing but a solution containing a culture of manipulated bacteria – that is, bacteria which have been genetically altered to carry out the function of digesting all the organic

> elements present in the water used. An engineer friend of
> mine told me that in some countries this product is placed
> in tanks in the basements of houses and blocks of flats. Is
> this comprehensible?

Yes, yes, it was comprehensible all right: they had simply put into practice something that already seemed obvious – in 1975! Way back then there were plenty of seminars and articles which showed that the sanitation technology engineers were using was obsolete, and that there was no way it could keep up with the pressure of demand. By the end of the century, so the argument went, twenty out of the twenty-five biggest conurbations in the world are going to be in what's left of the Third World, with sewerage systems covering only a few kilometres and scarcely any sewage treatment works.

Leptospirosis, cholera and typhus will spread endemically, and, while science is busy discovering the cures for cancer and for AIDS, millions of citizens of the marginalised countries will go right on dying of known and preventable diseases!

So when could this turnaround have happened? At what point had biology and engineering entered a vital partnership to solve this problem? How – and with what political commitment – had an innovative and cheaper technology finally succeeded in replacing one that people were used to employing, with all the weight of inertia and convenience behind it, however costly it was? Could the rivers, streams and lakes in the urban areas of Calcutta, Bombay and Shanghai have been restored to life? And could the streams in capital cities such as Teheran, Jakarta, Seoul and Delhi have been saved from being choked to death? Could sewage pollution have been eliminated in other urbanised areas such as Lagos, Cairo, Karachi and Manila? And could the water supply, degraded for so long, have been restored to good condition in major cities such as Beijing and Tianjin in China, or Lima, Mexico City and São Paulo in Latin America?

I went on reading the fax, and it seemed almost as if Titus had guessed what I was going to ask:

This technological innovation was not introduced rapidly; there was natural resistance to the change both from those who carried out engineering studies and designed projects and from the suppliers of products for sanitary installations. It was only introduced after the public outcry caused by the great epidemics of leptospirosis, typhus and cholera. If it hadn't been for this scare, which shook the World Health Organisation at the beginning of the century, and for the resulting flood of investments in science, technology and sanitation, we would still be holding academic discussions today on the possible existence of coprophagous bacteria and whether they might be controlled ... As for Turkey ... your curious prediction about its role in the world is not far off the mark, I think. Of course the "blue-turbaned son of Mahomet" arose, as we well know. But Nostradamus should be read seriously and not in such a literal fashion. There was destruction, but not in the form of a pitched battle in the Renaissance style. The core that was destroyed was not physical, not Rome, although the Catholic Church was affected; it was what one might call a virtual core – much more important, because what was changed was culture. And it is not by chance that Turkey played a central part in overcoming this conflict. But I shall have to return to this subject later; unfortunately I have to interrupt this fax – already so long – to go back to the teleconference. Such a pity! I confine myself to answering your query: no, the Istanbul committee and the teleconference initiated there are not about the situation in Turkey; they are about Cuba.

In solidarity and warmest friendship,

Titus

20:30

"Cuba – why Cuba? The institute where you work – what does it do?"
I scribbled down these questions, adding three more that I was
worried about: "Has a cure for AIDS been found yet? Will there be
problems with water shortages in the twenty-first century? Don't
forget to tell me what you wanted to say about the Internet!"

I had already stopped bothering with greetings and farewells in
these rapid-fire messages. While I was waiting for an answer I phoned
and ordered a pizza, as I expected and hoped that our diachronous
dialogue would go on into the night.

With my eyes glued to the Toshiba, I hadn't noticed it had
begun to rain noisily outside. A breath of moist air came up from the
earth and the grateful leaves trembled excitedly. It occurred to me
that Titus might not be able to lay his hands on some cuttings I had
filed regarding the grim prospects of world water shortages which
explained the reasoning behind my last question.

According to the World Meteorological Organisation the
volume of available drinking water has fallen steeply, particularly in
Latin America and in Europe. Although the populations of Africa, Asia
and Oceania will also suffer water shortages, the WMO believes that it
is on these two continents that the outlook is becoming most serious.
They say that there are three main reasons for this situation: the
practice of dumping chemicals, which get into the rivers and – even
more serious – reach the water-table, the growth of the urban
population, and bad handling of natural resources.

I assume, in fact, that water is going to become the most precious and certainly the scarcest natural resource in the next century. Setting aside the potential drinking water represented by ice and by the water-table, only 2.59 per cent of the total volume of water on the planet is drinkable: the rivers and lakes. And into the rivers and lakes we dump more and more polluting substances and sediments, while only 40 per cent of the water we use is returned to nature and recycled.

Demand, on the other hand, is shooting up. We consume 35 times more water today than we did three centuries ago, and consumption is rising at almost 8 per cent a year. The population of the globe was 1 billion in 1850 and is 5.5 billion today; estimates suggest 6 billion in 1998, 8.5 billion in the decade of 2020 and maybe 10 billion in the year 2050. Environmentalists believe that this is the maximum our planet could support. How can we guarantee the water we need?

Suddenly I remembered that Titus had said something about "water draining from the shower and sink being used for flushing the toilet". So at last something was being done about reusing and saving water! It is not easy to introduce the reuse of water in large countries where our culture is deeply imbued with a sense of the vastness of the space we live in, and we have a tendency, which arose spontaneously centuries ago, to see nature and its resources as a cornucopia of infinite capacity. Maybe I'm just being a cantankerous old man, but I think it's high time we incorporated the notion of saving, of economising our means and recycling our resources, into our cultural values.

We have to give up the childish game of hosing every last leaf all the way to the kerb instead of simply sweeping the pavement ... Well, some spendthrift habits could be changed just by increasing the water rates. But there's more serious work to do too, such as reducing the vast amount of water lost by seepage: in the São Paulo water supply network alone, up to 30 per cent of the drinking water – which has been treated at great expense, of course – is just being allowed to leak away! Would it be a pipe-dream to introduce obligatory grey water systems for domestic use? The water used for flushing the toilet, for watering the garden and for washing the car and hosing down the pavement wouldn't come from the water tank or the mains supply any more, but would be drained from showers and bathroom sinks via a secondary tank.

At the urban level we should be gearing up to use grey water for street cleaning and in industry too, and there's much more we could do. Rainwater could be collected in large underground channels and tanks, and as well as that we could maintain extensive green areas – linear parks along the bottom of every valley – to allow the wealth of rainwater to filter down to the water-table. And as most water is still used for agricultural purposes, more sophisticated irrigation technology should be compulsory.

I was turning all this over in my mind when the purring of the fax machine snapped me back to a state of alert once more.

Dear J. (May I use your forename? What is it?),

Just a few words regarding your concern about water shortages. I think the situation is worse than you predict. Thanks to the desertification of much of Africa, catastrophic famine spread from the area where it first appeared, in the so-called horn (Ethiopia, Somalia and Eritrea), to many other countries. Vast numbers were forced to emigrate, swelling the numbers of displaced peoples. For twenty years they lived in camps, only to end up being expelled and pushed out into other countries and other continents. As you can imagine, the arrival of great groups of African people in Europe and the Americas at the beginning of the century aggravated existing racism and provoked the violent incidents I think I mentioned earlier. And the immigrant population in Europe rose from 20 to 50 million in the course of twenty years ...

Water recycling is a reality today, but as usually happens it was started very late thanks to the resistance and inertia that hinder any pioneering proposal. But at least we have advanced quite a lot in irrigation technology. Today the world programme known as "Food by Satellite" (the proper name has an acronym impossible to memorise: UNSATMONFP – the United Nations Satellite Monitored Food Programme) links a sophisticated weather forecasting programme with regional and local irrigation controls, with partial mobilisation of water table resources and vast cisterns to collect any excess. I should explain that this programme also owed its success to the vast network of satellites in differentiated orbits whose equipment was adapted; their maintenance is a routine matter these days thanks to the ease with which they can be reached by small spaceships with autonomous energy supply and independent movement. What a difference there is between the vast array of innovations in the space sciences this century and the situation which, according to my notes, existed during the decades we call the "Cold War" (1960 to 1990), when a man set foot on the Moon as a symbolic act of victory during what was so inaptly called "Star Wars".

"Food by Satellite" was also possible because of the discovery and development of a series of food plants which required less water and which were genetically adapted to poor soils. And lastly we have Israeli technology to thank for all the sophistication of the irrigation systems used world-wide today. The nature of your questions throws light on an aspect of interest to my masterplus: it seems that the end of the last century was more tense and confused than I had imagined. You had not the slightest idea of what lay ahead! I shall come back to this later. Here it is almost midnight already, and if you are agreeable I do not intend to move from beside my fax machine. Let us continue with our exchange for as long as this fantastic opportunity for contact lasts! Think you not?

In his haste Titus even stopped signing off with his unfailing and odd "yours in solidarity". His reference to our ignorance of what lies ahead left me with the unpleasant feeling of being trapped inside a vast bubble – or maybe lost in space would be closer to the mark. What was it that we were failing to notice?

Anyway it was some consolation to me to know that Israel would continue with its strategy of supplying the world (and especially its neighbours) with inventions and technological advances instead of dramatic shows of force.

I sent Titus a quick message about this straight away, and he answered as follows:

Once it had given up the strategy of confrontation with the Arabs known as Palestinians, Israel, under the leadership of politicians of Sephardi descent – perhaps inspired by the felicitous coexistence of Arabs and Jews during the model Moorish occupation of the Iberian peninsula – formed an alliance with the new state that was created at last after the Gaza experiment. This state was entirely dependent for the first ten years of its existence on the Israelis themselves, thanks to the inter-Arab disputes which followed its establishment. I refrain deliberately from citing the name adopted for the new Palestinian state because of the idiotic and irrelevant disagreement about whether or not it would be proper to use the name "Palestine", which was the subject of impassioned debate at the United Nations for years.

But I cannot refrain from pointing out how Israel set an

example for political wisdom and long-term planning: they realised as early as the last century that shortages would make both water itself and low-cost alternative technologies into basic essentials, much in demand especially in the more promising of the slow-moving countries which were still struggling bravely to escape increasing poverty and yearning in vain to join the club of the fast-track countries. Israel made ready to become the main creator and exporter of these basic technologies, hence its current political strength of a very particular nature. How curious the way that the pioneering approach to social, scientific and technological areas of development strategy this century should have been characteristic of two such radically different countries at the same time: Israel and China! Both banking on technology, on science and on the whole of society playing its part, both subject to their demographical problems (of opposite kinds, of course) and both with decades of authoritarian, though not actually anti-democratic, government behind them ...

21:00

I was just wolfing my pizza and washing it down with a Coke, still firmly glued to the Toshiba, when another message came through from Titus, reminding me to answer his academic enquiries and pressing me with his much-repeated "think you not?" (which I finally identified as the common – and badly translated – German expression "glaubst du nicht?"). I was concerned to realise that in order to answer some of his questions about everyday life I was going to have to wait until the end of Carnival to get access to sources. The most I could lay hands on there and then in my study was the day's paper. I cut out a few articles and headlines and even some advertisements I thought were the most typical examples of the moment and sent them off, hoping that Titus' equipment was good enough for him to make use of the material.

I asked him to send me some newspaper clippings or advertisements and he answered my request quickly, but he explained that although his teleconference was finished he could only send whatever he happened to have on file for the documentation of his masterplus – hence, a Brazilian paper.

Advertising, he told me, had adapted to gradual changes in the press; newspapers didn't depend so much on advertising revenue any more but on the sale of information. Compared with turn-of-the-century publications, he wrote, the newspapers in his day and age were more like what we know today as weekly magazines – full of concise information with a lot of pictures. Longer texts and more in-depth writing now went into an infinite range of more or less specialised bulletins, newsletters and home journals which were distributed electronically to millions of subscribers.

Generally speaking we subscribe to a newsgroup with each person selecting and deciding on what news they want to receive for the period of their subscription. This news arrives as a package using various different support media: bulletins, electronic newspapers, television (by cable, of course), CD-MM and micro CDs. There are even electronic publications available by subscription, sent via satellite or cable, which provide specific summaries to order taken from the material circulated in other bulletins! Within this ensemble of information sources, the actual newspapers themselves function as a kind of "arouser"; they are quick and easy to read, with plenty of illustrations and a small number of well laid-out advertisements.

I remember how exciting I used to find the news-stands of my adolescence, with all those magazines on display – mixing headlines with naked girls, if I recall. Their equivalent these days would be the news nuclei: spacious booths equipped with all the available news displays, which we consult before reproducing them to take home on paper, tape, video or diskette.

Images are persuasive, doing away with the need for description and generating emotions, but they are not an efficient way of generating thought. Words are still essential for that. There was a Spaniard – called Giner if I remember rightly – who foresaw as early as the last century the way companies in the information business have become "information refineries" these days; they send everything electronically, aware of their responsibility to work fast since the speed with which information is provided forms the basis of their customer expectations.

But still, even though we order what we want to receive, personally I also enjoy the random surprise of getting information I did not ask for. And after being on the receiving end of an avalanche of startling images and bulletins of precise data, when we really want to understand what is going on we resort to an old habit: we sit down and talk things over with a friend ...

Titus sent me a clipping, which my black-and-white fax machine reproduced with the usual poor quality in a range of grey tones and black, in which an entire page was taken up by the photograph of a model dressed in lingerie rather different from the scanty underwear we are used to these days: instead of high-cut tanga briefs showing off her butt she was wearing something modest and ample, but to make up for that her young breasts – although they were being supported somehow, as far as I could make out from this poor reproduction – were bare and swathed in lace. At least I think it was lace! Across the photo appeared the incisive message: "**Don't waste energy! Put it to nobler uses**" And on one side there was a table showing the best energy source to choose for different uses: hydroelectric, nuclear, solar, hydrogen, biomass, reversible, heat exchange, natural gas, recycled, combustion, wind. A message of advice was stamped at the top of the advertisement: "Choose your supplier today and pay the lowest rates – summer, peak-time, inter-harvest or compensation".

There had certainly been a lot of changes: integrated multiple energy sources and multiple tariffs. But some things hadn't changed: the creative designers in our advertising industry were still just as talented and just as clever at using *double entendre*, the appeal of the erotic was still just as strong as ever in our culture and – last but not least – our models were still every bit as beautiful!

The other advertisement – also Brazilian – was more intriguing, however. It showed a couple with their arms around each other, gazing out of a train window at a dazzling nocturnal landscape. The advertising copy read: "Honeymoon deal – from Porto Alegre to São Luís for only B\$140". The advertisement was offering an overnight trip at the advantageous price of 140 Brazilian dollars between two cities almost 4000 kilometres apart! I was stunned; was Brazil going to have long-distance trains again? And that fast?

Titus shed light on this in his accompanying message:

> Brazil is criss-crossed by long rail links carrying the TGV, a Franco-Brazilian train financed by a Russian corporation with experience in the use of long-distance trains. For freight purposes they have connections with some of the principal waterways. As far as I can understand the TGV floats electromagnetically and travels thanks to an induction motor at an average speed of 350 km an hour.
>
> After a pilot project linking Rio de Janeiro and São Paulo, with only one stop, the first line in Brazil ran from the new Brazilian sea-port São Sebastião to Callao on the Pacific Ocean, crossing through Mato Grosso and Peru. As a matter of fact routes such as the Rio de Janeiro/Belém line or Porto Alegre to Bogotá, with stops in Recife, São Luís and Manaus, really changed the face of medium and long-distance transport.

So, I thought, it seems that the truck doesn't exactly have a brilliant future ahead of it ... and no wonder! Brazil is the only country in the world today to base all its long-distance freight transport on a low-capacity diesel-powered vehicle!

Titus added that in scanning his files he had noticed how the emergence of this new private rail network at the beginning of the century had also stimulated investment which enabled the foundation and growth of new cities, resulting in a denser network of urban centres in the area of the central Brazilian plateau, "the rich savannah

54

of Brazil's abundance", which would certainly be a step forward in our development. He continued:

> This question of transport in continental countries like yours has not always been approached in a rational manner, think you not? I believe, however, that the interlinked networks of railways, subways and waterways which we have today, together with the economic linkage between industrial complexes and airports, means that road freight can finally acquire a certain rationality.
>
> I refer to the hydrogen-powered autocontainers. Ever since Mazda, BMW and Mercedes launched the first hydrogen-powered cars at the end of the century there was naturally competition between manufacturers to use the new fuel (whose price was already comparable with the biomass fuels in vogue – ethanol and methanol). The oldest people, especially in the United States, still talk about the glorious days of cheap gasoline! Oh God! I can barely believe it, because we would never dream today of burning it in engines, this precious, non-renewable natural product so important in chemistry, for prostheses and for the manufacture of cosmetics and synthetics. Think you not?
>
> But here at the Kurz Institute (I forgot to tell you the name of the institute before) we cannot afford the luxury of feeling surprised; the history of wildly rising and falling petrol prices is much too important to an understanding of the last century. Perhaps you are knowledgeable about this?
>
> If so, as a matter of fact I would be grateful if you could correct me on the following version of events. The price was maintained artificially low (US$1) during the decades of imperialist action in the Middle East. After slight concessions, the first petrol crisis broke in 1973 when the price jumped from US$2 to US$10 a barrel. Do you

55

remember how the banks seized upon these sudden extraordinary profits, with a flood of petrodollars to the hitherto unheard-of total of US$250 billion, made available instantly at low interest rates to the so-called "developing" countries whose domestic savings were insufficient so that they were thirsty for capital? There was a mad rush into debt fuelled by great enthusiasm on both sides, think you not?

The second crisis came in 1979, when prices reached US$40 a barrel under the control of the petrol producers' cartel OPEC. The petrodollar loans encouraged by the big banks rose to half a trillion US dollars. And so the immense foreign debt of the poor countries (that today we call "slow countries") came about, and the foundations were laid for what the economists of our Institute call "indebted industrialisation". Is this interpretation correct, do you think?

History records that ten years later the price collapsed to US$10 a barrel owing to internal divisions between members of the producers' cartel and an indiscriminate increase in production. So the poor petrol-producing countries could not pay the debts they had incurred in the expectation of exports, and those which, like Brazil, had no petrol to export, were already finding themselves bogged down in debt thanks to their rash, indiscriminate use of petrodollars. And so, imperceptibly at first, there began the phase of "indebted de-industrialisation". Correct?

But according to the criteria adopted in our institute the financial and economic problems of the end of the century – such as the devaluation of the American dollar, the rise and then the fall of their interest rate, the creation of the ECU (the European unit of credit, a new accounting currency) and finally the world recession – cannot be attributed solely to speculation on petrol prices.

The speed at which capital was now moving around the

world, the play of the stock markets, the US budget deficit and foreign debt and the increasingly autonomous and extra- or supra-national structure of the transnational companies all showed that the play of the economy depended less and less on the commodities and raw materials which had governed political pressures for centuries. Forgive my lecturing tone, dear J., it is a professional failing! When I get enthusiastic about a historical episode I cannot stop ...

And as you cannot interrupt me, I shall go on! While I am on the subject of the pressures that govern politics, I must emphasise how dramatic it was that the so-called cheap labour of the backward (slow-moving) countries came to be a factor of decreasing importance in what used to be called the international division of labour. The rapid technological progress of the fast-moving countries (some people call them the central countries, and at the beginning of the century some authors called them the Pragmatic Capitalism countries) caused such profound changes in the processes of production that at the beginning of the century the slow countries suddenly found themselves left out of competition on the international market altogether, either because their labour force quickly proved unsuitable, or because they did not have the necessary capital for rapid technological change. The slow countries were left with nothing to offer in order to compete: neither a cheap and suitable labour force nor abundant, sought-after raw materials. But the central countries, already prey to the trauma of structural unemployment, were prevented from taking advantage of this situation by the fact that the very marginalisation of the slow countries became yet another de-stabilising element for them! I would even say that the slow countries as a whole ceased to be irrelevant – as they had been for a decade – and became a catalyst of sorts by precipitating, however indirectly, the gradual but radical transformation of the system of production!

22:15

The end of this fax came as a shock. Was the system of production –
the very capitalist system in which developing countries so ardently
wish to play a major role – going to collapse? Seneca's words sounded
truer than ever now: "There's no fair wind for those who know not
where they want to go."

Still, I had no right to be surprised. As early as ten years ago
there were a lot of us who pleaded with every successive government
to pay more attention to education and to culture in the broader sense
of the word, in order to generate the capacity for work and creativity
needed to keep up with the rapidly changing demands of the system of
production. With almost 19 per cent of the adult population of 15 or
over illiterate in 1990, with almost 15 per cent absenteeism at primary
school level, with technical education lacking all prestige, with higher
education courses more concerned with preserving the privileges of
the professional bodies – hell, here in Brazil we're actually reducing
our potential, and the same is true for so many developing countries.
And, as the above makes clear, we're gradually writing ourselves off
from any chance of competing on the world stage.

Besides which, we're still looking at any form of investment in
science, technology, art and culture as if it were some kind of
superfluous luxury instead of realising how strategically important
these things are: without ideas and creativity we're hardly likely to be
able to make real plans for development. And if we can't plan for it,
how on earth can we ever achieve it?

This is stuff I'm forever turning over and over in my mind, but I'd better not bore you with it now. Back to the diachronous dialogue with Titus!

22:45

By now I was dying to call time-out so that I could think all this over at leisure and set down a whole lot of questions about what Titus called the "gradual but radical transformation of the system of production". Instead, almost unwillingly, I returned to his fax about autocontainers and finished reading it:

> This is a goods vehicle made up of a mechanical traction unit and an extendable chassis on which one or more modules designed to contain freight are placed. But of course you must know the system; it became imperative to construct these containers in modular form at the beginning of the century in order to make them suitable for the robot handling systems introduced to cope with the increasing volume of goods moving through the transfer stations that had to be built at the entrance to major cities and at sea, river and air ports.
>
> During the last few decades freight and urban transport have undergone considerable changes, as you must have foreseen. Apart from the new fuels, the emergence of the public car and the new popularity of the high-speed train, there are various other factors which have led to a decline in the use of the vehicle people used to call a "truck":

integrated systems have been developed for transporting large loads and goods in bulk by river, and the new industrial districts are built close to airports (these districts are never large like they used to be, thanks to the increasingly decentralised reindustrialisation of the major cities). These days the automobile industry has to make a real effort to refine autocontainer design if it wants to keep up its sales figures, and the transfer stations – which are completely robot-operated and assembled on a modular design in accordance with the volume of goods they are forecast to handle – were introduced thanks to the encouragement of the vehicle manufacturers themselves in order to guarantee the continuity of freight transport by motor vehicle. I have spent many pleasant moments visiting such transfer stations with my son, simply to delight in the ballet of the containers as the robots transfer them onto urban goods vehicles according to their destination. These goods vehicles are generally electric, since the new batteries invented for this purpose came onto the market. Oh God, only a few days ago I was showing my son some relatively recent photographs from my files in which urban goods were slowly and laboriously unloaded in residential streets by porters who carried them upon their backs, despite the protests of the residents!

I imagine that the changes in urban transport met with some resistance, but it must have been utterly swept aside by the serious incidents of extended atmospheric pollution which paralysed the capital cities of Tokyo and Seoul for a fortnight at the beginning of the century. The penultimate United Nations Environment Conference (not the last one in 2012 but the one before, held in Rio de Janeiro in 1992) put the discussion of the urban environment centre-stage after twenty years in which environmentalist debate had been concerned only with attacks against the natural

> environment outside urban areas. This awareness of the
> need to reclaim and improve the environment in which 90
> per cent of the population were living certainly carried
> weight in the reorientation of vehicle use and technology.
> This is the common interpretation of the facts these days; is
> it correct?

Titus ended his message there. I was amazed he had no complaints about how late it was! For him it must already be the early hours of Sunday morning. I sent him a fax to break the tension and give him fresh heart:

> It's two o'clock in the morning, my friend, you must be tired
> and hungry. Have you eaten anything? All the best!

And Titus answered:

> Yes, I am ... but I have a kitchenette with a freezer and an
> oven, so I took a moment to eat something. A pizza and some
> cola. Let us press on with our dialogue, which is already
> lasting longer than we could have hoped, think you not?

So I went on turning things over in my mind, wondering what would become of our cities with their desperate need to make up a huge and growing shortfall in housing, sanitation, transport and landscaping. In the year 2000 the world's urban population will be 2.85 billion. It was almost 2.5 billion in 1980, when the predictions were that it would increase by 4 billion in less than fifty years in the Third World cities alone! Would we have reached a total of 6.5 billion people living in cities by about the year 2030? I couldn't help wondering anxiously what life would be like in the capital cities of the countries Titus called "slow", given the dramatic economic marginalisation which my correspondent so dryly described. Suddenly I noticed that Titus had mentioned the name of the institute where he worked: Kurz. Probably named after the writer who was calling the Third World countries "post-catastrophic" in 1991!

And I wondered: were we heading for the catastrophe of marginalisation and the irremediable advance of poverty in our part of the world?

23:30

Dear J.,

While I was waiting for your last message I whiled away a few moments watching television, admiring the annual show you Brazilians put on to mark the occasion of Carnival. Oh God! For us here in the Community it remains a mystery how you manage to stage such scenes of joy and sensuality in spite of having to live with such difficult problems to resolve. It really gives the impression that it is all genuine and spontaneous! Do people really enjoy capering about and displaying themselves? Is it really thanks to a collective effort that each group is so well turned out? Is the all-pervading sensuality really an authentic trait of Afro-Brazilian culture? Friends who have been there assure me that we are the ones who have got it all wrong: we fail to understand simple things because we lack sensuality. I answer them in some annoyance that such sweeping statements merely over generalise.

I remember that, according to the data I have retrieved for my masterplus, during the last century this show used to be financed by the illicit system of a lottery based on animals. Is it not so? It occurs to me that this might be a good moment to tell you something about the rise and fall of the world of illicit activities which was perhaps the greatest historical landmark of the first decades of this century.

All over the world there are still vestiges of the parallel economy which was so marked during this period. The terrible events of the turn of the century, with their stupidity and their mindless violence, opened up a vast terrain for the expansion of two types of criminal activity: neoterrorism and the post-mafia organisations. Of course the rising numbers of assassinations – almost all of which occurred in the prosperous fast-moving countries, full of pockets of marginalised people – arose out of a lack of capacity to govern (in spite of economic recovery), out of the seething cultural brew of growing masses of poor people of immigrant origin cut off from their roots, and out of the vast numbers of young people with no prospect of work. In this context it was easy for any psychopath to come up with a rallying cry offering an "aim in life", a cause to fight for based on death and destruction. For every targeted assassination or mass-murder there sprang up a wave of groups, movements and factions claiming responsibility: those on the margins of society were desperately seeking a way of making themselves visible, of existing. There was a tense, frantic need to look for and find a collective aim in life. Urban society was divided into veritable tribes, each with its own banner to fight under!

The result was predictable: here in the Community, for example, we all lived in a state of siege imprisoned behind our own defences – whether physical and electronic barriers, or a mental defence of ingenuous and desperate alienation which made us ignore everyday dangers. It was in this phobic environment that I was brought up in my parents' home, as they spent every waking moment activating some defence and prevention system or other.

The urban landscape of our cities was made up of bars and

traps, barriers, sensors, alarms and watch-towers, with those ubiquitous electronic spying eyes recording everything. You can still see this landscape today, not just in the old central countries but also in the big capital cities which, for a while, were a part of the "islands of welfare that form an archipelago in an ocean of marginality" (I am quoting from one of your own papers which I retrieved a few days before my first contact with you: as you can see, I am well documented!).

It is no great consolation to say that writers, poets and artists had expressed their presentiments about all this during the preceding decades. Humanity was rushing headlong towards its worst crisis with the inevitability of a Greek tragedy, think you not? A death foretold, as that twentieth-century writer García Marquez would have said ...

Neoterrorism took its name from the fact that instead of making use of crude car-bombs it had unfortunately become more sophisticated: on the one hand kidnapping as a form of extortion had positively become customary in the capital cities – cynics regarded it as "the fastest way of diluting the uneven concentration of income" – and on the other the practice of invading computers and databases had become a common form of blackmail and source of threats. For a handsome price, the electronic terrorists would supply the solution to recover a vital file or get rid of a hidden virus blocking programmes of major social impact whose clock was ticking away – even though institutions defended themselves with equal electronic sophistication, and had their files continuously scanned by electronic sentinels!

The invasion of the Internet and its struggle for freedom was a significant turning-point and a major defeat for neoterrorism. At the turn of the century (and perhaps already in your time) the Internet was an informational network that spelled freedom from governmental controls and from owner-domination. A century after Kropotkin, here we had an anarchistic free political instrument! Of course both bureaucrats and the mafia dreamed of manipulating it. It would take too long to describe all the successive struggles, the electronic invasions and the defences thrown up against them over a period of three years. Radical changes were made, including the change of all access codes, and this period ended with the defeat and imprisonment of the criminal gang behind the attacks.

But while the psychopaths were acting on a raison d'être of their own, this was a world run by organisations: it was the mafias that really built up a flourishing parallel economy. The days were long gone when the Sicilian organisation, later emulated in the United States by immigrants of other origins, used to confine itself to trafficking in illegal alcohol, gambling and prostitution. The days of the primitive drugs trade and selling protection were over too.

What did spring up and flourish at the beginning of the century was a sophisticated illegal network operating in parallel with the productive activities; a tax evasion system par excellence with its representatives, its media, its statistics, its market expansion strategy for the products it concentrated on at first: drugs, child prostitution (supposedly AIDS-free), pornography, gambling and especially arms. And later: tourism, the hotel trade, air transport, professional sport, cinema, pop music, television, advertising, the press, fashion ... Later the mafias went on to establish art galleries and finally museums whose collections were made up of works acquired at rock-bottom prices at auctions and from dealers: a cultural way of laundering money ...

If you would like more details about this I could send you the minutes of legal proceedings instituted during the 2010s in Germany, England, Japan, Korea and Switzerland (which acted only after everything had blown up in other countries), constituting a sort of moral revival of the legal system's anti-mafia struggles in Italy at the end of the century.

But before these legal proceedings a very disturbing phenomenon occurred in the institutions governing the economies of the central countries. The quantities of money that had to be laundered in order to enter legal circulation were so large that they ended up generating their own banks, and their financing and investments began to compete with those of the traditional investors.

Little by little the internal logic of currency circulation, which was still in force at the beginning of the century, rendered the frontier between these two worlds tenuous and finally non-existent. The need to launder money simply disappeared: governments required investment to prolong the life of the system of production and to finance what was left of an already impossible welfare state (because of the rising cost of supporting their unemployed), and they gradually adapted their laws to this de facto situation! It would be interesting for my masterplus to know to what point you estimate that this split in financial activity has already progressed in your country.

The activity of the numerous mafia organisations gradually gained recognition and even some degree of respectability. Then it began to be called the "parallel economy", and there was no lack of neo-liberal ideologues who were so anxious to explain that everything was rosy in the capitalist world that they constructed ingenious theories about this type of "heavily deregulated" economy. My parents always used to tell me how difficult it was to bring up children at the beginning of the century! Except perhaps for those of orthodox religious convictions – of whatever faith – whose fundamentalism erected barriers against an awkward reality. It became a daily gamble, striving against all the odds just to make out what was correct, just and honest.

They used to tell me that all sense of civic morality degenerated to such an extent during the last decade of the last century that whenThe European newspaper ran a survey on it they found that a significant proportion of those interviewed in France, Belgium, Portugal, Spain, Germany and other Community countries considered it acceptable to evade taxes, avoid paying on public transport, claim state benefits to which they were not entitled, accept bribes or knowingly buy stolen goods!

It is not by chance that in this century people began to live in small, tightly knit groups, with clusters of families in small private streets or in isolated spots looking for privacy and intimacy, turning in upon themselves, embracing introspection or seeking to limit their communications in an attempt to escape the alienation of mass existence, the mass media, the populist appeal to the masses, generalisation and vulgarisation. People were forced into a constant round of negotiations and concessions by their very awareness of absurdly conflicting realities – conflicts that they could only reconcile either by denigrating their own perceptions or by embracing a state of alienation. In the central fast countries an incensed and divided society was living imprisoned by the double economy, while in the peripheral slow countries this situation was aggravated by the constant frustration of marginalisation.

In Brazil you were really quite lucky, according to the data I have been retrieving, partly because your economy did not match the levels and dynamism of the central countries. It did not compare with the only two to have a foreign trade surplus (Japan and Germany) or with the USA, whose economy underwent a spectacular, sudden and disastrous collapse and was rebuilt, or with the surprising Cuba, or with those countries which made a last-ditch attempt to make the compete-to-export blueprint a reality: the economies of Singapore, Taiwan, Korea and China's increasingly prosperous export bases led by Hong Kong.

Though ill-prepared for competition and blind, for a long time, to the opportunities represented by the transnationalisation of the economy, Brazil was saved by diversity and biomass, think you not? (Oh God, you cannot know yet!). I would even say (please do not take offence) that according to the data and studies I have retrieved, the Brazilian economy at the end of the century was like nothing so much as the spectacle of Carnival: each one dancing alone, moving to a mind-numbing rhythm – the rhythm of modernisation – all dressed in cheap but spectacular fantastical costumes, all eager to display themselves, all progressing with great slowness – because they are really walking sideways! And yet, in spite of everything, there is some hidden logic in accordance with which they do nevertheless still move ahead.

I have just re-read the last paragraph. I leave it in only because you must understand that this is only a fax, not a book defending any theses. Besides which, it is three o'clock in the morning and the functioning of my intellect is getting a little precarious.

Brazil was also saved by another characteristic: its ethnic tolerance. We all know that racial discrimination existed there too, think you not? But nothing like that which was seen in other parts of the world at the turn of the century! I refer principally to the Community, to Russia, to the former Yugoslavia, to the USA-Canada Consortium and to Japan.

71

The feelings of tribalism which were present in every city, especially those overrun by immigrants, gave rise to violent intolerance at the end of the century, which manifested itself at best in divisive irony and humour, at worst, in the homicidal and genocidal outbreaks of violence of the period. People invented the most impracticable forms of nationalism, the better to hate and persecute others. Social groups were excluded and marginalised on the basis of national traditions, often newly invented and quite artificial but nevertheless held up as symbolic and "eternal", in a tremendous confusion between culture and state, without anyone ever reflecting on the fact that the whole paraphernalia of national symbols dated only from the nineteenth century! It was as if the mere existence of the hated and feared "other" justified the affirmation of some nationality or other. I will come back to this matter in a later fax, as it is crucial to an understanding of the period I am researching.

In Brazil you were culturally accustomed to the migration of peoples, to racial miscegenation, to religious syncretism, to a pragmatic flexibility in everyday morality. Everything which we despised as lack of national identity, looking down from the "heights" of our historical culture here in the Community, shone out like a beacon of humanity in a world darkened by intolerance; it was a unique ethical value, and it pointed the way towards our salvation. It took twenty years for us to realise how harmful these nationalist passions were, when in reality they only served to justify the most reactionary of tribal movements based on the hatred of anybody and everybody who happened to belong to a different tribe! We had not realised before that we were regressing from a society in which the individual could shine to a tribalist society in which the only thing to be seen was the mob, the stupid, primitive collectivity!

Oh God, forgive me for not yet having given a direct answer
to your questions about city life today. They must wait until a
little later. I ended up getting carried away talking about the
parallel economy, the apartheid and tribalism of the end of
the century, just because they were events of vital
importance. But I will return to your enquiries. Send me
your comments straight away ...

In solidarity,

Titus

0:10

Oh God!, as Titus would exclaim in his literal translation of the stock expression "Ach Gott!". So we are going to have to face the cynicism of an official decision to incorporate this "parallel economy" run by the mafia, the drugs traffickers and related interests, are we?! This means that when that scandal broke about the Karachi-based bank which specialised in subsidising terrorism, laundering drug money and financing terrorist activities consistent with mafia interests, what we were witnessing was only a trailer for the financial institutionalisation of crime on a world scale, a "superior" way for the drugs traffic to operate within the economy?

On the other hand Titus kept harping on about the "terrible years of intolerance and violence" but he kept putting off going into detail. I wanted to know where and who, how did it start and how did it finish? And besides, I found it hard to believe that the Brazilian knack for adapting to a life of co-existence – the mulatto solution – was going to become an envied and admired virtue, a world rarity, at the turn of the century. We always considered it a facet of the lordly, elitist style we liked to put on; tolerant because superior ... So our culture has ended up generating an unexpected value quite involuntarily!

I wrote a hurried fax setting down a series of direct questions, rushing to put my ideas in order.

And this was when the first mishap occurred: I couldn't get the message to go through! I repeated the number, which I had already learned by heart even though it was so complicated – but in vain. I let a little time go by, with my imagination running wild about what might be going on at the other end: Titus had hung up, there was a fault on the line, some problem with the satellite, no-one in ... what if the magic and our diachronous dialogue had come to an abrupt end together?

A cold shiver ran up and down my spine and I found myself numb and sweating. Not now! It wasn't fair, that contact should be broken just when I was so close to learning something so crucial ... Could it be the mafia intercepting our exchange? I put this hypothesis aside as too paranoid; I'm not given to conspiracy theories. Some dislocation across the temporal pole? How could I turn back the clock, how could I restore contact?

I made myself a solemn promise: if I ever managed to contact Titus again I would order my thoughts and give priority to basic questions and answers; I had to accept there was a risk that we might get cut off for good, and quite possibly very soon. I re-read all our correspondence and asked myself what ought to get top priority. Everyday life in Titus' day? The dangers of intolerance at the turn of the century? The content of the Havana Conference in which he was taking part? The "current" studies of the Kurz Institute? A cure for AIDS?

Or perhaps I should be less selfish and should try above all else to answer the questions which my friend, with his unfailing "solidarity", needed to help him with the writing of his thesis?

To gain time I started here; I began preparing the bibliographical information he had asked me for, using the books I had in my study which made reference to the subjects we had mentioned so that he could use his computer to retrieve the works in question:

| Castells, Manuel | *The Informational City* (Blackwell, 1989) |
| Drucker, Peter | *The New Realities* (Butterworth & Heinemann, 1989) |

Erdman, Paul	*What's Next?*
	(Doubleday, 1988)
Furtado, Celso	*Brasil, a construção interrompida*
	(Paz e Terra, 1992)
Galbraith, John Kenneth	*The Age of Uncertainty*
	(Houghton Mifflin, 1977)
Hawking, Stephen	*A Brief History of Time*
	(Bantam Books, 1988)
Kurz, Robert	*Der Kollaps der Modernisierung*
	(Eichborn & Co. , Frankfurt, 1989) Titus
	must know this one, of course ...
Lades, I.	*The Next 40 Years*
United Nations	*1992 Population Fund Report*
Sachs, Ignacy	*Stratégies de transition vers le XXI siècle*
	(Studio Nobel/FUNDAP, 1993)
Sassen, Saskia	*The Global City*
	(Princeton University Press, 1991)
Toffler, Alvin	*Powershift*
	(Record, 1990)

This done, I turned back to my Toshiba, gave it a hard stare, sighed and – feeling like a basketball player about to take a vital free throw – I tried again. And this time ... it worked! It worked! The fax went through, OK, OK, OK! Titus and I were still in contact!

So now to work: I prepared a series of numbered questions and sent them at once, and was already about to send another series when I heard the tinkling of the fax once more. After the usual long wait its mechanism whirred into life and it spat out its white sheet which I began to read upside-down in my anxiety; it seemed that Titus too was getting nervous about the danger that contact between us might get broken off ...

1:00

My dear J.,

What a shock! My God! I tried to send you a brief message and it did not go through! Did you also experience a similar difficulty? It would appear that we cannot draw out our instructive diachronous dialogue too far; we must sum up and conclude. It seems that as time goes by we must be reaching the farthest limits within which contact is possible. I have re-read everything we have sent one another and I will try to sum up as best I can in order to answer your questions. It would be advantageous if you could take this opportunity to do the same thing also, think you not? By the way, I thank you for the bibliographical recommendations. Some I have already consulted in non-Brazilian editions; I shall try to retrieve those which I have not yet consulted.

Some explanations to begin with: the "masterplus" is the theoretical work required after a period of academic retraining which we normally undergo periodically, each in our own area of professional activity, after working outside the university for a number of years. It is difficult to obtain employment without this periodical retraining and the corresponding final paper, as competition is extremely fierce. This constant to-and-fro between academic life and a life of non-theoretical, productive professional activity has considerably speeded up development in the applied sciences as well as increasing productivity. The universities have also benefited from the return of professionals coming from the "outside world" of production.

With regard to the institute to which I belong, it was created by a number of followers of Robert Kurz, who must have been your contemporary. He would have hated being enthroned in an institute, since he always fought against institutions, but a government decided to pay tribute to him: although his book on the end of the system of production was criticised for its reductionism (which irritated specialists but delighted the lay reader) he nevertheless provoked a polemic which provided a wealth of material for academic studies. Which is why this institute, at some point, received the name of Kurz.

But my role here has been confined to helping organise the Havana Conference and, now, to putting in order the documents resulting from it.

I shall try to explain what this major conference was about, mentioning only the main topics initially; if we find we have enough time for our exchange I will return to this subject and relate each topic in more detail.

The conference took shape at the beginning of the 2010s after the Babel Lynchings. Why "Babel"? I shall return to the subject in detail, but for now it is enough to note that these bloody and sadistic events, which occurred in precisely that city which we all considered the capital of culture and modernity, provoked an astonishing outcry which aroused humanity's conscience once and for all. So why "Babel"? Because every language under the sun could be heard amidst the cries of the poor immigrants, their homes turned into burning traps, when they tried to escape and were brutally clubbed to death at their own doors! It was an indescribable horror, broadcast on TV over and over again around the world ...

After twenty years of aggressive intolerance, of ethnic hatreds, of a parallel economy which supported terrorism and violence, the world had taken on an appearance that historians described as "mediaeval": it was an archipelago of city-states inhabited by the few millions who had all the income, surrounded by oceans of pariahs, the permanently unemployed, the outcasts, the wretched and the misinformed, all striving in vain to find a way into those islands of well-being. No "First World", no "Third World", just "citizens" on one side and the "marginalised" on the other – all belonging to some kind of organisation or other, of course – co-existing with all the tension of a veritable state of siege dominated by fear!

The conference brought together all the leading decision-makers: the national governments, the Joint Chamber of the Transnational Companies, the Agency for the Control of Weapons of Mass Destruction, the Ecumenical Synod of Religious Authorities, the Banking Compensation Fund (in which the institutions of the national and parallel economies merged and were interconnected). Over a period of a few years, social movements for a new humanism emerged in almost every country; many of them adopted the slogan "Humanity First" which indicated society's main concern and showed that they were against setting too much store by pragmatism in decision-making. These movements were represented in Cuba too. I will return to this subject if we have time ...

Finally the conference met in Havana for a month, in the Autumn of 2018. Some journalists tried to suggest that the choice of year was an attempt to commemorate the centenary of a revolution which happened in Russia. But it was pure coincidence, though we could speculate on an academic level that the kind of "new humanism" everyone talks about these days constituted a new utopia whose power to mobilise was similar to that of the utopia called socialist which brought down the Russian czarist regime in 1917.

There is also another parallel that might be drawn with something that happened a century before. Today, as then, the music of the Russian composers crosses national frontiers: the new Russian operas are wonderful (and even popular), with their magnificent interactive multimedia events in which the audience takes part. And as far as the visual arts and design are concerned, there are many Russians who devote themselves to activities linked with the socio-political movement of new humanism – like Malevich and Tatlin a hundred years earlier, when the Suprematist and Constructivist schools allied themselves so movingly with the popular redemption promised by the socialist utopia in Russia. History presents us with apparent repetitions and cycles seem to repeat themselves down the spiral of the years but, I repeat, in the case of the date chosen for the Havana Conference this was mere coincidence.

A good many intellectuals, scientists, ecodevelopmentalists, ombudsmen and information and communications technology specialists took part in the conference, as well as the federations of marginalised people everywhere and representatives of the people's autonomous city chambers.

Its introductory assessment of the situation – drawn up beforehand as always, by various committees over the previous five years – included observations on the following: a) the division of the world into fast-moving and slow regions; b) the confederate structure of the transnational companies as they operate on the global economy; c) the progressive self-destruction of the system of production, which was being replaced increasingly rapidly owing to the disruptive effect of the parallel economy on the traditional

82

economy of production; d) the chaotic sea of information in which we were drowning, a result of excessive secrecy in the way information and communications technology was being used to organise information flow; e) the manifest popular rejection of all violence arising out of the lynchings incident; f) the gradual transformation of a number of fundamentalist religious movements into movements of spiritual solidarity and the consequent higher regard for ethical values in society; g) the apparent replacement in some countries – particularly in the case of some successful experiments in city-states – of the "market" economy (so-called before the days of strong state intervention in the economy of various countries at the beginning of the century) by what these days we tentatively call the "social economy". The speakers quoted from Drucker in one of the debates when they noted the existence of four interdependent economic agents: the individual nations, the united regions (such as the European Union, the Islamic League or NAFTA), the flow of capital/credit/investments and the transnational companies, all of which were unsettled for some time by the activities of the parallel economy. I mention this author because the conference paid tribute to him in its preparatory work as being the first author to identify the Transitional Trauma as a discontinuity in the course of history. Besides, the conference also included specialist meetings, one of which dealt with another subject Drucker had studied forty years before: the urgent need to consolidate the many existing specific agreements into a transnational law to replace our outdated international law.

An Indian participant whose name I cannot now recall said that we were currently witnessing the "final triumph of Marx", an author who already seemed old-fashioned last century after the dismantling of the Soviet Union. This writer, quoting from our namesake Kurz' polemical book, pointed out that the advent of regimes based on social economy confirmed the truth of Marx' analyses regarding commodity fetishism. The title of this lecture was "The fall of the Berlin Wall marked the end of Capitalism". I cite this example because it may be of some relevance to all of you. For us this is all history now, and the only people to get excited about it are professionals like me who study the past in order to understand the present.

In fact the rules of the old capitalist game simply could not be kept going any longer – not by all the stratagems of the mafia parallel economy nor even by the transnational companies which, in order to ensure their own survival, began to set up some interesting experiments in which the production of goods was combined with the social organisation of various communities on a scale of sufficient significance to sow the seeds of the new economic regime, to which some give the name of "social economy system of production", or "social economy" for short.

In this system the working day is shorter, allowing several shifts and thus a larger number of jobs. Having a lot of free time gave rise to a wide range of leisure activities, retraining and study and encouraged physical fitness and spontaneous creativity. And to think that this "otium cum dignitate", as the Ancient Romans put it, was generated by the structural unemployment of the Trauma years!

Right up to the beginning of the century, during the rise of state economic intervention, it would have been hard to predict that the most lasting experiments in social economy (besides the mushrooming of non-state public institutions) would be generated by the transnational companies. Today's theoreticians try to find the logic behind these events. A posteriori, as always ...

I get the impression that I have summed up the content of the conference rather too briefly. Besides, it has to be said that it was the fruit of five years of work on the part of numerous committees. And that this work was possible only because the banks of both the parallel and traditional economies had reached an agreement that the time had come to overcome the deadlock and the decline of the system of production and to look for a safe and gradual way of changing it. Working in these tense conditions, a variety of propositions raised by different social organisations ended up comprising the agenda of the conference. Everyone came together in favour of the consolidation of tolerance, of solidarity, of coexistence between cultures whose differences should be respected and valued. In sum: it was the glorification of peaceful coexistence in diversity!

It would be a great pleasure to detail all the ups and downs of the five years of preparation, worthy of an adventure story in themselves, such as the conflicts of interests, the intrigues and the pressures brought to bear on the members of the committees. I took advantage of the personal experience of having taken part in organising the conference in order to put my masterplus on the agenda as a minuscule but relevant topic, since in it I study and describe the particular experience of one country, with a majority of marginalised people in its population, during the transformation which resulted in the new humanism.

If we should have time I will return to the subjects of the conference. I should also like to tell you something about the "Long Live Difference" movement and to answer some of your other enquiries as soon as possible. Keep up the contact! Good-bye for the moment.

In solidarity,

Titus

1:50

When I had read this fax I sat for a few minutes unable to move. I knew I had to seize the opportunity of contacting Titus to raise more questions and to send him some of the information he asked for, but on the other hand this wealth of important new material demanded careful thought.

How and why would this high level of state intervention in the economies of various countries come about at the beginning of the next century when, with the collapse of the Soviet Union, everything today suggests a strong tendency towards liberalisation and privatisation in economies throughout the world? And what was the political and administrative structure of the city-states he mentioned?

For transnational companies to have an increased level of effect on the global economy would require state involvement in the economy, certainly – but not of the traditional kind; it would focus more on the non-economic aspects of development (the "external factors" referred to by economists) and on the problems of national sovereignty in the face of the intense globalisation of economic decisions.

This strengthening of state influence might be provoked by the difficulty of financing the social services, aggravated by the marginalisation of the unemployed. But, on the other hand, the new level of state presence which Titus mentioned might have more to do with the disturbance caused by the parallel economy – the economy of the mafia and the drugs trade – when it reached levels such as to have a significant impact, forcing the state to intervene in order to preserve the national financial system and defend each national currency, hence defending its own sovereignty.

It is also easy to imagine that the growing social tension arising out of poverty and marginalisation, out of apartheid and violence (and the emergence of new private terrorist armies), could lead the state to police life in a given country in a number of ways such that its influence would be towards centralisation – while it would be difficult, on the other hand, to imagine that the ingenuous *laissez-faire* of old could survive under such circumstances. In this case, the current vogue for liberalism and deregulation will prove significant only in today's specific circumstances – a short-term corrective to the excessive bureaucracy and corporatism to which the influence of many states has been reduced.

As for this modern version of the city-states of ancient Greece, it put me in mind of some debates on urban planning in which various participants have defended the thesis that we should "forget about the so-called great national problems and concentrate on solving local problems". In fact we would be a great deal further forward if we could achieve good environmental levels in our cities: sewage treated, sound and air pollution controlled, reasonable public transport, differentiated road networks, adequate shade and drainage enabling the existence of parks and trees. Cities might have more vitality if we could get a sensible mixture of functions in land use, mixing housing, work and services and leisure with a balanced density throughout the city, opportunities for access to information, education and cultural development for every citizen and the creation of employment opportunities to meet the needs generated by the city itself. Here and now!

Even though it wouldn't cover all the problems of national development, I thought to myself, a policy centred on urban development would certainly prove popular and would be a powerful motivating force in public opinion. A policy like this could be the breakthrough that opens the way to the more basic, structural political changes Brazil so badly needs.

And I remembered how, in his book *The Informational City*, Castells observes the growing importance of the computerised flow of intangibles in our everyday lives and expresses his concern for the loss of importance of physical place, of urban spaces; how he suggests that levels of local power, which protects and generates the urban space, should be increased in order to prevent the citizens – powerless and ignorant – from remaining the pawns of this computerised flow. In this way the city would have the condition of a state, allowing its citizens to exercise a more direct influence on activities.

Afraid that I would take up too much time in transmitting this train of thought, I limited myself to a short message asking the following questions:

Is Fukuyama an important author in the Institute's studies? Does the state continue to exercise a predominant role in the economy of nations? Does the city-state have a specific legal and institutional structure?

And then a new problem stopped me dead: the fax machine failed to transmit my message and curtly informed me of an "**Error**". I checked that Titus' long number was correct. I tried again several times, with a feeling of discouragement and sadness beginning to well up in my heart. But after fifteen minutes of anxious attempts the mechanism accepted my note and my words flew (by what paths?) into Titus' hands!

Minutes later I received his answer: we were still in contact!

2:30

Dear J.,

I have been having difficulty transmitting messages to you! I am afraid that our time is running out. I am hurrying to send you more information while I can and to ask you to clarify some matters, knowing that at any moment we could be cut off forever! You cannot imagine how much this upsets me ...There is no time for farewells, but I want you to know that I feel so very much solidarity with you!

To answer some of your questions: no, of course I cannot read or write Portuguese ... I write in my own language. I think I have already mentioned that it is the fax machine's translator modem which changes the language. The Kurz Institute currently devotes itself principally to contributing to the global network of information on experiments in social economy. Instead of preparing theoretical studies the institute concentrates more on case studies and on disseminating them among the city-states, taking its example from the pragmatism and the concern with placing intellectual effort at the service of immediate social change which characterised the last works of the author from whom we take our name. By the way, there are few cases in which these city-states take institutional form; their existence is de facto, since urbanisation this century has led to most of the population of the world living either in the city-states themselves or in the urbanised regions of which they form the poles.

I am sorry but I cannot place Fukuyama. I do not have access to the Anthropos XX database at the moment. Who is he?

I shall give a brief response to one of your other questions: in my masterplus I am trying to describe and comment on the roots of the new humanism in the context of the marginalised countries. And I use Brazil just as a good example. Basically, I would like to be able to show that this humanism arises out of a highly varied mosaic of contributions, out of an interactive process of acculturation.

Perhaps there is enough time to write a little more about the important Havana conference. Try to send me some information about the question of ethics in politics, an interesting subject (vital at the beginning of the century) which was touched upon lightly for a short time in the Brazilian press — dealing with corruption at the highest level, according to the archives I have retrieved. Until circumstances delete our dialogue,

yours in solidarity,

Titus

I had barely read to the end of this fax when the next, even shorter message came through:

You asked me whether we have sustainable development in my time! I have difficulty understanding the question: can there be such a thing as "development" which is not sustainable? Are you perhaps referring to some kind of economic growth, and not to what I understand as

92

development?

On the other hand you are right to be concerned about the situation of poverty and the lack of equipment and sanitation in the large cities of what was formerly called the Third World. In the long term there was a better distribution of the population: after the explosive growth of some of the capital cities in the slow countries, particularly the Asian cities, various factors eventually contributed to evening out density within the space of these nations and creating a better urban network – factors such as epidemics, the high cost of living and new production opportunities in less urbanised regions associated with the growing use of biomass. According to the annual report of the United Nations in 2010 the situation was reasonably satisfactory in this regard even in the slowest countries. Nevertheless, sanitation shortfalls continued and were overcome only thanks to technological breakthroughs such as those I mentioned earlier. Progress is not usually linear, or, as I like to explain to my children, "the electric lightbulb was not invented by perfecting the wax candle" ...

I was still digesting this information when my fax machine shrilled out its brief warning and produced yet another message:

It seems to me that our faxes can still transmit when they do not take so long, that is when the message is shorter. It is a theory like any other ... So I am answering some of your questions in small doses before trying to expand on the results of the conference. In earlier faxes which I have now re-read you asked about the foreign debt of the slow countries. The debt situation was one of great perversity. According to the data I have retrieved, in 1983 the countries

93

belonging to the so-called Third World owed US$644 billion; by 1990 they had paid out US$670 billion in partial payment and in debt servicing – and yet their debt had increased to US$950 billion! The poor countries were exporting capital to the rich ... and still they remained in debt.

The situation got worse during the last decade of the century, with the rapid entry of the Eastern European countries into the debtors' club while under the delusion that they were making their triumphal entry into the select circle of the rich countries! On the contrary, their future faithfully mirrored the difficult situation of the Third World countries – though with the important difference that they had good distribution of income, good health services and a solid basic education system. Generally speaking, the foreign debt of what you still call the Third World was only gradually resolved at the beginning of the century thanks to bilateral agreements between the debtor countries and conglomerates of transnational companies with some support, in certain countries, from agents of the parallel economy. The TNs and the mafia (my friends do not like it when I use this word so bluntly, recalling the illegitimate origins of the parallel economy) became aware that keeping these countries in a state of poverty or stagnation greatly reduced the economic space they needed for their legitimate economic activities. And so they set up negotiations which allowed the system of production to be kept going for a little longer.

And then there came another "quick-fax". Maybe Titus really had found a way for us to continue our diachronous dialogue. I began preparing my messages in homoeopathic doses too. In this fax my friend expressed his surprise at my question about the formal expression with which he ended his messages – the words "in solidarity":

If you ask me such a question perhaps I did not describe emphatically enough how the social fabric was torn apart at the end of the century! Xenophobia, ethnic hatred, exclusion and apartheid predominated until as late as the year 2010; we lived in a state of constant tension and insecurity, whether in Berlin, Tokyo, Lagos, Paris, Chicago or Rio de Janeiro. At the same time, though, a reaction was germinating in citizens and society: human solidarity was an expression of salvation, of redemption from social chaos – which is why it has become part of our culture in so many ways. Not everyone would agree with me, but I would even go so far as to say that this human solidarity was the basis of the new humanism discussed at the conference.

While I was scribbling down my comments the fax machine – which was now working and collaborating tirelessly – rewarded me with yet another quick-fax dangling head-downwards waiting to be read. It was Titus' answer to a question on AIDS:

As for the questions you raised about health problems, the statistics show an increase in longevity: life expectancy went up over the last hundred years from 55 to 70 and finally to 90 in the fast-moving countries. This was owing primarily to preventative medicine, made possible thanks to the deciphering of the genetic code and the identification of the genes contained in the human genome.

But it was a result of other factors, too. There were considerable improvements in our diet thanks to the increased range of natural foodstuffs made available on the market; there were new techniques for treatment at the cellular level and there were micrometrical transplants in order to correct functional defects; new scientific knowledge was brought together from various different origins, ranging from microsurgery to applied genetics; there were new ideas about pre-natal traumas and their corresponding psychological phenomena; there were natural analgesics and new discoveries in pharmaceutics resulting from good use of the biomass and of the drugs derived from the human immune system about which fundamental discoveries were made at the turn of the century.

And of course it was thanks to these discoveries that we were first able to bring the AIDS epidemic under control and finally developed the vaccine which saved us from it at the beginning of the century after an appalling number of deaths. In the wake of this research, cancer – the "harmful programmed proliferation of cells" – was also restricted when its psychosomatic and immunological aetiology was discovered. I even think – according to my consultation of the Anthropos XX – that someone with the same surname as yourself was interested in this matter and developed a thesis on cellular memory, the relationship between psyche and soma and the programming of what happens to each person starting with a basic pre-natal pattern.

Although these terrible illnesses have been brought under control now, the same cannot be said of the endemics that are the mark of poverty: people are still dying of cholera, of leptospirosis, of starvation, of diarrhoea, and now they are dying of tuberculosis again. People are dying of the so-called avoidable diseases .... shame on us! However, this is not a scientific but a political problem, and although the Havana conference drew up a public timetable for change, and although it brought together various different sectors and oriented them towards dealing with this problem, it could not wave a magic wand and transform the widely differing realities of all the nations in the world overnight.

4:15

The minutes passed. After a brief jingling noise that came to nothing, my office grew heavy with an oppressive silence. In Titus' country it must be broad daylight. Could he have fallen asleep? Sadly I resigned myself to the fact that we must have lost contact. It had to come sooner or later ... Mentally I began to put this wealth of new information in order, wondering how on earth I could tell people about something I had learned in such an unaccountable fashion. If all this were to be seen as mere speculation, as some theory that I had thought up for myself, I would be robbing my unknown friend of his rights as the real author. But if I were to disclose everything just as it had actually happened, as a diachronous dialogue, I would certainly run the risk of being called a lunatic or a liar! Perhaps I should publish what I had learned as fiction, leaving it up to the readers to judge the objectivity of my story for themselves ...

It had already gone four o'clock in the morning when my Toshiba suddenly jingled into life again – but this time it didn't stop short: once more I heard its mechanical purring and watched the long white sheet slowly pour out until at last it curled across the carpet in baroque swirls. Astonished at its length I tore it off and read:

My dear J.,

I was most frustrated when I found that I could no longer

97

send the brief messages with which we were still managing to maintain contact. Oh God! (I have not had time to correct this expression on my modem translator and accept its suggestion – "Oh shit!" – although I suspect that this is not an exact enough translation): I even went so far as to strike my machine in anger and it went dead! As a result I had to find an unlocked room with another machine to use instead! You can imagine my distress as I ran up and down the corridors finding only code-locked doors, and all the time afraid that you might be sending me messages that could not be received!

Settled at my desk once more I took another of those pills of plant origin which help me to stay awake and reasonably lucid (they are essential if you work for an international organisation or spend your time attending conferences). Now I am trying to send a long message, contrary to my earlier "theory" about short messages. I want to send you information once and for all on some unresolved matters relating to the Havana conference.

Certain ethical values came through strongly in the conference debates: those which formed the basis for humanity to establish concepts on which there was some consensus through the painful process of growing towards maturity at the turn of the century which some called the "Transitional Trauma". I shall limit myself to listing them: intimacy, privacy, romanticism, affection, the extension of the family, regarding waste as an anti-social sin, holding the individual and the human figure in high regard, a certain (I would almost say Dionysian) hedonism, giving prestige to artistic creativity – these are the values which make sense in our day. Hence we speak of a new humanism, in the classical sense used by the historians.

Perhaps in reaction against the violence of the Transitional

Trauma, people repeated an expression taken from the
Jewish philosopher Hillel so often that it became a cliché:
"Do not do unto others what you would not have others do
unto you. That is the law. All else stems from this ..."
Beautiful, isn't it?

These days there is a real gut need for reconstruction,
starting with the most intimate aspect of the self, the most
individual, the very core of one's being. This is jealously
guarded, perhaps in reaction against the fashion for
"opening one's heart", the exhibitionism and exposure
which characterised art and life at the end of the last
century. Immersed in the rich life of the metropolis, we
strive to recover the indispensable space of privacy and
silence – indispensable for introspection and particularly for
personal re-creation which implies the digestion and
assimilation of the constant avalanche of information which
has speeded up the rhythm of daily life to an excessive
degree. These new values probably emerged in response to
the constant mental shifts and displacements demanded of
us in our daily lives. Hence – to take a droll example – the
success of these one- or two-person booths that have sprung
up in our parks and other urban spaces in which we can
enjoy silence and gratify our senses with images such as
abstracts or secluded landscapes or even dream images,
which we seek to reproduce so as to bury them in our
memories once and for all. We can construct surreal images
in these booths – perhaps the cultural product of videoclips
from the end of the century – or simply enjoy the pleasure of
silence or listen to sounds from nature or to music. What a
difference from the violent and pornographic material that
used to spill out of every VR headset in the past! I admit that
of all our new cultural habits, some of which may be
transitory, the current fashion for the extended family will
probably not survive. The provisional marriage – an

institution which acts as a preliminary to a possible marriage contract – was succeeded by the custom of bringing another male or female friend to live as part of the family. Although this custom succeeded in eliminating the feeling of guilt associated with extramarital affairs (and at the beginning of the century it also reduced the risks of transmitting venereal diseases) it seems likely that frequent jealousy, some gender confusion and occasional possessiveness will probably prevent this system from becoming universal.

Jealousy, envy ... acceptance, rejection ... being on the inside or being left out ... are there any human sentiments more lasting or that make us suffer longer or more deeply than these? Most of the last century was characterised by social movements demanding "equality" for women: equal rights, equal pay, equal opportunities. These days we understand that, although these movements and the results they obtained were right and just, and that they brought social progress and enabled women to fulfil roles that benefited all of us, there was something unresolved inherent in this new position of women.

They were now subject to tremendous stress, divided as they were between work, professional career, politics, pregnancy and maternity, small children, the avalanche of information, as well as the increased number of problems in human relationships, with their helpings of emotions, affections, passions.

So it is no surprise that in this century women have fought successfully for their "exclusive" characteristics to be recognised and held in higher regard: their capacity to gestate and to mother, their greater competence in looking after the house, the right to exercise their power of seduction (also known as femininity). "Down with equality, long live difference" proclaimed the provocative posters to be seen in every language and in every city, and whose content also influenced the resolutions of the conference.

To sum up, there are two slogans which symbolise the cultural changes to be seen in our everyday lives – from the relationships between people to the production of consumer goods, from cultural production to fashion: "Humanity first" and "Vive la difference", with the content described above.

But to move on to another subject, the Havana conference also brought some immediate results. Of these I shall mention one: the immediate creation of the World Development Fund. Although some had fought for this over twenty years ago, it was established in operational form only in 2019. The fund is made up of 1/1000 per cent of the Gross World Product – which to give you some idea is equivalent in 1990 statistics to almost US$20 billion. Taken in addition to the cancellation of the foreign debt at the beginning of the century, as I mentioned earlier, over the next few years this annual distribution of funds noticeably reduced the accumulated factors separating slow- and fast-moving countries – at least in the case of those SCs which showed competence in the use of their own appropriate resources.

So the fund proved to be an important instrument of support during this period of transition when the system of social economy was introduced throughout the world; thanks to its humanist rationality (a common expression today among politically correct intellectuals), this system should prevent a return to those terrible and unjust distances that used to exist until a short time ago between the countries of the First and Third Worlds (to use the nomenclature to which you are accustomed). Think you not?

With regard to competence in the use of resources, it is worth recalling what happened in the last few decades in the area of so-called "human resources". Those who, like me, delight in the Romantic literature of the nineteenth century, will remember the predominant presence of a professional category which does not exist today: that of the domestic servant, the professional category which was then the most numerous! It makes you remember how things change ... In a similar fashion, although the category of "industrial workers" – the numerous proletariat which was so characteristic of the twentieth century – does still exist, they now comprise less than 10 per cent of the workforce in the most important fast-moving countries: Japan, China, the Community and the United States. In these countries the predominant category is that of the "employees" of some organisation or other, whether public or private, frequently small or medium-sized. The socio-economic structure in China is exemplary and most interesting in this regard.

Most of the slow countries, however, lag behind in the training and re-training of staff – that is, in education – and this has kept their economies stuck at a backward stage: that of exporting raw materials (which are valued less and less), with manufacture replacing imports! The exceptions were that group of countries which you used to call the Asian Tigers and China, with the great "cultural revolution" which occurred in that country in the last decade of the century (of course I do not refer to the short period of fanatical retreat during the last century which was – absurdly – given this same name).

The SCs are still paying a high price for the absence or shortage of policies which would develop basic healthcare and education, eliminate illiteracy, support the flourishing of the sciences and of art – in short, all the forms of acquiring knowledge. They are also paying dearly for their lack of planning, that is, the failure to deliberate about the future. As a result, history tells us how even at the end of the century many SCs launched themselves with enormous effort down the road of old-style industrialisation, still dreaming about a system of production which was already being called into question and about an international division of labour which no longer existed! ...

And they undertook this great effort at a time when conditions already made it possible to recognise the transnationalisation of the economy and understand what it meant, when there were already visible signs that the existing system of production was undergoing a radical change, when it was already clear to see how difficult it would be for Brazil and other SCs to undertake a process of modernisation capable of restoring them to a position in a globalised economy. Do you not consider that this question of modernisation, taken as the hub of all development strategies, has been a terrible mystification? Or could it have been a mere mistake, innocent though serious?

Even though things have changed this century, as you may well have foreseen already, it seems to me that it has never yet been explained sufficiently why you Brazilians kept up for so long the dangerous and growing tension between "the Masters and the Slaves" (to quote the well-known author Freire, whom we hold in high regard here at the institute). What was the motive, psychological or cultural, that prevented your country from seeing the way the world was going so that you might take advantage of the current and make a leap forward, making the most of the country's vast existing resources: its population with their characteristic tolerance and creativity, its space, its biomass! ...

Tell me, if we still have time, what it was that paralysed your country at the end of the century and blinded it to the real changes, what stopped it from seeing the opportunities it might have in a world thrown into turmoil by apartheid, by violence, by the radical changes in the system of production?

103

What was the motive for Brazil's scepticism, for its immobility? Do not tell me that you were waiting for the return of Dom Sebastião! (Do you know what I mean? Forgive me, this is the cultural historian coming out in me).

In the end, what was it that prevented Brazil from drawing up a national plan to drag it out of its backwardness-with-a-look-of-modernity? Was there some lack of passion, lucidity, political will? Or was this a cultural trait of which people were still barely aware? Keep trying to communicate with me: it would be important for my masterplus for me to understand this turn-of-the-century Brazilian phenomenon.

I wonder: what will the people of Brazzzzzzzzzzzzzz

And here Titus' fax broke off. His question hung in the air, unfinished.

This was the last message I received, and although I tried desperately over and over again I was unable to send any more messages myself. Something told me that the situation had changed for good: the diachronous dialogue was over.

The day was just beginning. From outside there came a dull thump as the Sunday paper hit the ground, thrown by the delivery boy. I asked myself ironically what great news it could possibly bring me, after what had happened to me since yesterday!

Somewhere in the space-time of the universe I had just lost touch with a friend. My friend Titus, with his eternal solidarity ... Another phrase from Francis Bacon came into my mind: *Time, like space, has its deserts and its solitudes* ... Tired, I sat for a long time gazing at the last line of the fax:

I wonder: what will the people of Brazzzzzzzzzzzzzz